Methods for the Numerical Solution of Partial Differential Equations

Modern Analytic
and Computational
Methods *in* Science
and Mathematics

A GROUP OF MONOGRAPHS
AND ADVANCED TEXTBOOKS

Richard Bellman, EDITOR
University of Southern California

Published

1. R. E. Bellman, R. E. Kalaba, and Marcia C. Prestrud, Invariant Imbedding and Radiative Transfer in Slabs of Finite Thickness, 1963

2. R. E. Bellman, Harriet H. Kagiwada, R. E. Kalaba, and Marcia C. Prestrud, Invariant Imbedding and Time-Dependent Transport Processes, 1964

3. R. E. Bellman and R. E. Kalaba, Quasilinearization and Nonlinear Boundary-Value Problems, 1965

4. R. E. Bellman, R. E. Kalaba, and Jo Ann Lockett, Numerical Inversion of the Laplace Transform: Applications to Biology, Economics, Engineering, and Physics, 1966

5. S. G. Mikhlin and K. L. Smolitskiy, Approximate Methods for Solution of Differential and Integral Equations, 1967

6. R. N. Adams and E. D. Denman, Wave Propagation and Turbulent Media, 1966

7. R. L. Stratonovich, Conditional Markov Processes and Their Application to the Theory of Optimal Control, 1968

8. A. G. Ivakhnenko and V. G. Lapa, Cybernetics and Forecasting Techniques, 1967

9. G. A. Chebotarev, Analytical and Numerical Methods of Celestial Mechanics, 1967

10. S. F. Feshchenko, N. I. Shkil', and L. D. Nikolenko, Asymptopic Methods in the Theory of Linear Differential Equations, 1967

11. A. G. Butkovskiy, Distributed Control Systems, 1969

12. R. E. Larson, State Increment Dynamic Programming, 1968

13. J. Kowalik and M. R. Osborne, Methods for Unconstrained Optimization Problems, 1968

14. S. J. Yakowitz, Mathematics of Adaptive Control Processes, 1969

15. S. K. Srinivasan, Stochastic Theory and Cascade Processes, 1969

16. D. U. von Rosenberg, Methods for the Numerical Solution of Partial Differential Equations, 1969

17. R. B. Banerji, Theory of Problem Solving: An Approach to Artificial Intelligence, 1969

18. R. Lattès and J.-L. Lions, The Method of Quasi-Reversibility: Applications to Partial Differential Equations. Translated from the French edition and edited by Richard Bellman, 1969

19. D. G. B. Edelen, Nonlocal Variations and Local Invariance of Fields, 1969

20. J. R. Radbill and G. A. McCue, Quasilinearization and Nonlinear Problems in Fluid and Orbital Mechanics, 1970

21. W. Squire, Integration for Engineers and Scientists, 1970

22. T. Parthasarathy and T. E. S. Raghavan, Some Topics in Two-Person Games, 1971

23. T. Hacker, Flight Stability and Control, 1970

24. D. H. Jacobson and D. Q. Mayne, Differential Dynamic Programming, 1970

25. H. Mine and S. Osaki, Markovian Decision Processes, 1970

26. W. Sierpiński, 250 Problems in Elementary Number Theory, 1970

27. E. D. Denman, Coupled Modes in Plasmas, Elastic Media, and Parametric Amplifiers, 1970

28. F. H. Northover, Applied Diffraction Theory, 1971

29. G. A. Phillipson, Identification of Distributed Systems, 1971

30. D. H. Moore, Heaviside Operational Calculus: An Elementary Foundation, 1971

32. V. F. Demyanov and A. M. Rubinov, Approximate Methods in Optimization Problems, 1970

33. S. K. Srinivasan and R. Vasudevan, Introduction to Random Differential Equations and Their Applications, 1971

34. C. J. Mode, Multitype Branching Processes: Theory and Applications, 1971

37. W. T. Tutte, Introduction to the Theory of Matroids, 1971

In Preparation

31. S. M. Roberts and J. S. Shipman, Two-Point Boundary Value Problems: Shooting Methods

35. R. Tomović and M. Vukobratović, General Sensitivity Theory

36. J. G. Krzyż, Problems in Complex Variable Theory

38. B. W. Rust and W. R. Burrus, Mathematical Programming and the Numerical Solution of Linear Equations

39. N. Buras, Scientific Allocation of Water Resources: Water Resources Development and Utilization—A Rational Approach

Methods for the Numerical Solution of Partial Differential Equations

by

Dale U. von Rosenberg
Department of Chemical Engineering
Tulane University, New Orleans, Louisiana

American Elsevier Publishing Company, Inc.
New York 1969

AMERICAN ELSEVIER PUBLISHING COMPANY, INC.
52 Vanderbilt Avenue, New York, N.Y. 10017

ELSEVIER PUBLISHING COMPANY, LTD.
Barking, Essex, England

ELSEVIER PUBLISHING COMPANY
335 Jan Van Galenstraat, P.O. Box 211
Amsterdam, The Netherlands

International Standard Book Number 0-444-00049-6

Library of Congress Card Number 69-13069

First Printing, 1969
Second Printing, with corrections, March, 1971

Printed in the United States of America

CONTENTS

CHAPTER THREE

Linear Hyperbolic Partial Differential Equations

CHAPTER FOUR

Alternate Forms of Coefficient Matrices

CHAPTER FIVE

Nonlinear Parabolic Equations

CHAPTER SIX

Nonlinear Hyperbolic Equations

CHAPTER SEVEN

Nonlinear Boundary Conditions

CHAPTER EIGHT

Elliptic Equations and Parabolic Equations in Two and Three Space Dimensions

CHAPTER NINE

Other Types of Equations

PREFACE

During the last ten years engineers and scientists in all fields have used partial differential equations to describe an increasingly large number of their problems. These equations are of little value, however, if no solution to them is available. Only a few of them have analytical solutions, and partial differential equations cannot be solved on analog computers. Therefore, numerical solution of these equations using high speed digital computers is the only recourse. At the present time, there is no book available which describes these numerical methods so that the technical worker can learn how to use them to solve his problems. The book was written to fill this need.

A number of very efficient numerical methods have been developed in recent years, many of which are not complicated to use, although a knowledge of them is not widespread. In fact, it is only in the petroleum production industry that these methods have been used extensively for any period of time. Since the literature of the petroleum production industry is not widely read, there is a great need for the dissemination of knowledge of these methods.

The use of numerical solution methods is not difficult to learn, and no mathematical background beyond second-year calculus is required. In fact, it is more important for the student of this book to have a good understanding of various engineering problems and of the use of partial differential equations to describe them than it is for him to have a sophisticated mathematical background. This book is written so that a senior undergraduate or first-year graduate student in engineering or science can learn to use these methods in a single semester course, and so that an engineer in industry can learn them by self-study.

The methods are described on physical problems that the student will have studied in previous courses. They are introduced on simple problems and then extended to more complex ones. Emphasis is placed on the use of discrete grid points, the representation of derivatives by finite difference ratios, and the consequent replacement of the differential equation by a set of finite difference equations. Efficient methods for the solution of the resulting set of equations are given, and five solution algorithms are presented. A discussion of the solution of unsteady, one-dimensional hyperbolic and parabolic equations is presented in detail. An extensive discussion of the treatment of nonlinearities and of various boundary conditions is included. The solution of two-dimensional, unsteady conduction problems is also discussed in detail, and methods for the solution of more complicated problems are indicated.

DALE U. VON ROSENBERG

New Orleans, Louisiana
April 1969

Chapter One

LINEAR ORDINARY DIFFERENTIAL EQUATIONS

1. INTRODUCTION

Many of the differential equations which result from engineering problems cannot be readily solved by analytical methods. Consequently, a knowledge of the methods of obtaining numerical solutions of differential equations is important to the modern engineer.

A number of techniques have been developed for the numerical solution of ordinary differential equations. Most of these, such as the Euler method, the Milne method, and the Runge-Kutta method, are well known and are described in texts on numerical analysis. However, the more efficient methods for numerical solution of partial differential equations have been developed more recently with the advent of the high-speed digital computer. Consequently, these methods have not been described so that an engineer can learn how to use them. This text was written in an attempt to so explain these methods.

The more familiar methods for solving ordinary equations will not be discussed in this text. In fact, the purpose of the first chapter is to prepare the reader for the subsequent chapters on the solution of partial differential equations, since many of the necessary concepts can be described more readily with ordinary equations.

2. DIFFERENTIAL EQUATION CONSIDERED

These methods are most readily learned by studying their application to a specific differential equation and its accompanying boundary conditions. A numerical solution is always obtained for the differential equation with specific boundary conditions. These, of course, describe some physical problem. Most engineers closely associate the equations they wish to solve with some physical problem. Therefore, frequent reference will be made in this book to a physical problem described by the equations under consideration.

Let us consider the differential equation

$$\frac{d^2u}{dx^2} + \frac{2}{q+x}\frac{du}{dx} - \frac{2p}{q+x}u = 0 \tag{1-1}$$

This equation may be used to describe heat conduction in an uninsulated, tapered rod which is thin enough that a one-dimensional analysis can be used. The length variable, x, and the temperature variable, u, are dimensionless and normalized. The constants q and p are dimensionless combinations of parameters which describe the geometry and heat transfer characteristics of the rod. They are defined as

$$q = \frac{D_0}{fL}$$

and

$$p = \frac{hL}{k}\sqrt{1 + \frac{4}{f^2}}$$

where L is the length of the rod.
h is the coefficient of heat transfer between the rod and its surroundings.
k is the thermal conductivity of the rod.
f and D_0 define the diameter, D, of the rod by $D = D_0 + fLx$.

The simplest boundary conditions to use with this equation are specifications of the temperature at each end. The conditions to be used are

$$u(0) = 0 \tag{1-2a}$$

$$u(1) = 1 \tag{1-2b}$$

3. DISCRETE VARIABLES AND FINITE DIFFERENCES

The analytical solution of these equations specifies u as a function of x, where both x and u are continuous variables. To obtain a numerical solution, one replaces these continuous variables with discrete variables. The relations between these discrete variables are finite difference equations, and it is these finite difference equations which are solved numerically on a digital computer. We must obtain relations between finite difference ratios and derivatives to convert the differential equation to a set of finite difference equations.

To obtain these relations, let us review a relationship for continuous variables. Let u be a continuous function of x as described by the curve shown in Figure 1-1. Two adjacent values of x are separated by a differential length, dx; that is, the two points on the curve corresponding to the adjacent values of x are allowed to approach each other so that the distance between them approaches zero as a limit. In this case, the values of u at these two points are related by

$$u(x + dx) = u(x) + \frac{du}{dx}(x) \cdot dx \tag{1-3}$$

For discrete variables the distance between two adjacent points, Δx, cannot approach zero and must remain finite. In this case, equation (1-3) with dx replaced by Δx is not exact; the left side of the equation differs from the right side by the vertical line segment between the curve and its tangent

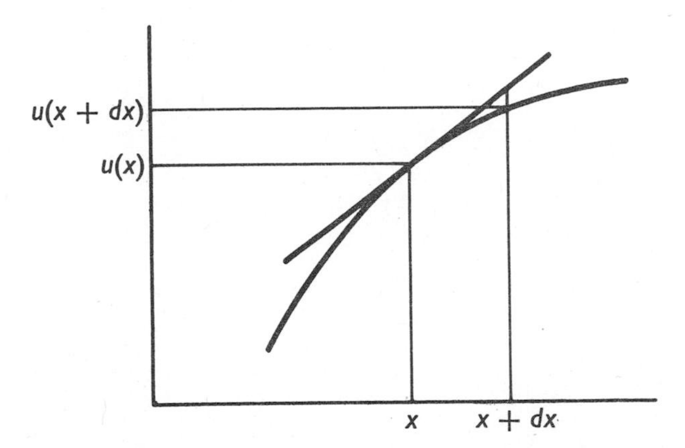

Figure 1-1. Derivative of a function.

at the point $(x + \Delta x)$. When adjacent points are separated by a finite difference, Δx, the exact relation is a Taylor series. This relation is

$$u(x + \Delta x) = u(x) + \frac{du}{dx}(x) \cdot \Delta x + \frac{d^2u}{dx^2}(x) \cdot \frac{(\Delta x)^2}{2!} + \frac{d^3u}{dx^3}(x) \cdot \frac{(\Delta x)^3}{3!} + \cdots + \tag{1-4}$$

4. NOTATION FOR DISCRETE VARIABLES

It is convenient to use a special notation for discrete variables. The discrete variable is defined at a finite number of points spaced equally in the interval between 0 and 1, as shown in Figure 1-2. The increment between the points is the finite difference, Δx. The value of the discrete variable at each of these points is denoted by the subscript i and is defined as

$$x_i = i(\Delta x) \tag{1-5}$$

The index i takes on integral values from 0 to R, where R is the total number of increments in the complete interval between 0 and 1. Some useful relations between values of the independent variable at adjacent points are

$$x_{i+1} = x_i + \Delta x \tag{1-6a}$$

$$x_{i-1} = x_i - \Delta x \tag{1-6b}$$

The notation used to refer to a value of the dependent variable which corresponds to a given value of the independent variable is the same as that used

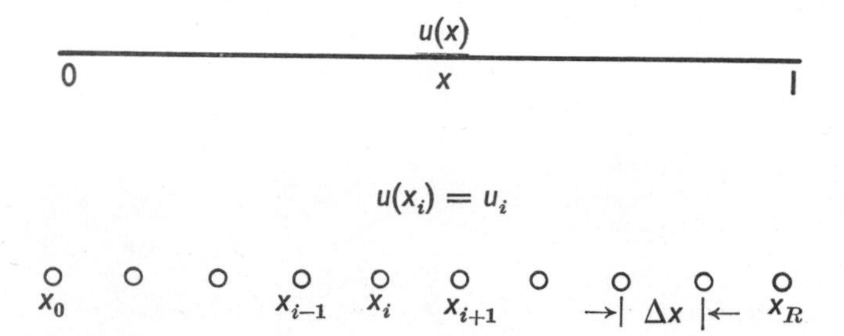

Figure 1-2. Discrete and continuous variables.

for continuous variables—namely, $u(x_i)$. It is more convenient, oftentimes, to refer to this value simply as u_i.

In the nomenclature given above, the Taylor series appears as

$$u_{i+1} = u_i + \left(\frac{du}{dx}\right)_i \Delta x + \left(\frac{d^2u}{dx^2}\right)_i \frac{(\Delta x)^2}{2!} + \left(\frac{d^3u}{dx^3}\right)_i \frac{(\Delta x)^3}{3!} + \left(\frac{d^4u}{dx^4}\right)_i \frac{(\Delta x)^4}{4!} + \cdots + \tag{1-4a}$$

In a similar manner, the value of u at x_{i-1} is

$$u_{i-1} = u_i - \left(\frac{du}{dx}\right)_i \Delta x + \left(\frac{d^2u}{dx^2}\right)_i \frac{(\Delta x)^2}{2!} - \left(\frac{d^3u}{dx^3}\right)_i \frac{(\Delta x)^3}{3!} + \left(\frac{d^4u}{dx^4}\right)_i \frac{(\Delta x)^4}{4!} - \cdots - \tag{1-4b}$$

Finite difference analogs to the first and second derivatives can be obtained from equations (1-4a) and (1-4b).

5. ANALOG FOR THE FIRST DERIVATIVE

When equation (1-4a) is solved for the first derivative, it takes the form

$$\left(\frac{du}{dx}\right)_i = \frac{u_{i+1} - u_i}{\Delta x} - \left(\frac{d^2u}{dx^2}\right)_i \frac{\Delta x}{2!} - \left(\frac{d^3u}{dx^3}\right)_i \frac{(\Delta x)^2}{3!} - \cdots - \tag{1-4c}$$

The first term on the right side of the equation is a finite difference analog to the first derivative. This term contains values of the dependent variable at two adjacent points and the increment in the independent variable. The error in using this analog is of the order of the first term which is truncated $(d^2u/dx^2)_i(\Delta x/2!)$. This term contains Δx, so the truncation error is said to be first order, and the analog is said to be first-order correct.

6. ANALOG FOR THE SECOND DERIVATIVE

A finite difference analog to the second derivative is obtained by adding equations (1-4a) and (1-4b). The result is

$$u_{i+1} + u_{i-1} = 2u_i + \left(\frac{d^2u}{dx^2}\right)_i (\Delta x)^2 + 2\left(\frac{d^4u}{dx^4}\right)_i \frac{(\Delta x)^4}{4!} + \cdots + \tag{1-7}$$

When this equation is written explicitly for the second derivative, it becomes

$$\left(\frac{d^2u}{dx^2}\right)_i = \frac{u_{i+1} - 2u_i + u_{i-1}}{(\Delta x)^2} - \left(\frac{d^4u}{dx^4}\right)_i \frac{(\Delta x)^2}{12} - \cdots - \tag{1-7a}$$

The first term on the right side of equation (1-7a) is a finite difference analog of the second derivative. It is obtained by truncating the series after this term and is second-order correct, since the first term dropped contains $(\Delta x)^2$. To approximate the derivative to a given degree of precision, one must use a smaller increment for a first-order-correct analog than for a second-order-correct analog. Consequently, a second-order-correct analog for the first derivative is desired.

7. IMPROVED ANALOG FOR THE FIRST DERIVATIVE

Such an analog can be obtained by subtracting equation (1-4b) from equation (1-4a). The resulting equation, written explicitly for the first derivative, is

$$\left(\frac{du}{dx}\right)_i = \frac{u_{i+1} - u_{i-1}}{2(\Delta x)} - \left(\frac{d^3u}{dx^3}\right)_i \frac{(\Delta x)^2}{6} - \cdots - \tag{1-8}$$

The first term on the right side of equation (1-8) is the desired second-order-correct finite difference analog to the first derivative, since the first term to be truncated contains $(\Delta x)^2$.

8. GEOMETRICAL INTERPRETATION OF FINITE DIFFERENCES

Before proceeding to find the complete finite difference equations to be used, let us look further at the analogs obtained. As shown in Figure 1-3,

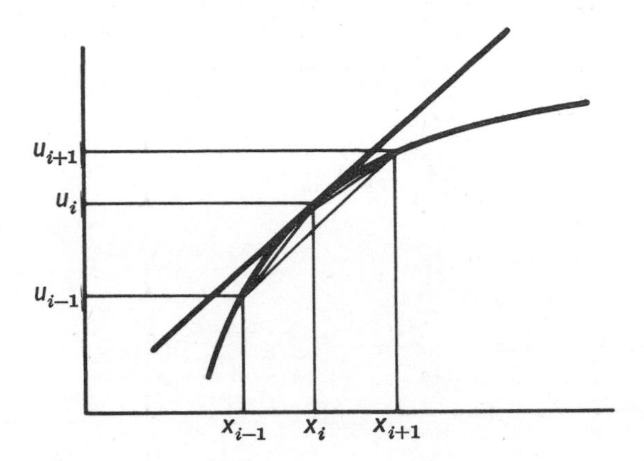

Figure 1-3. Derivative and finite differences.

$(du/dx)_i$ is the slope of tangent to the curve at the point x_i. The finite difference analog to the first derivative given by equation (1-4c) is the slope of the chord between the points (x_{i+1}, u_{i+1}) and (x_i, u_i). The analog given by equation (1-8) is the slope of the chord between the points (x_{i+1}, u_{i+1}) and (x_{i-1}, u_{i-1}). It can be seen from Figure 1-3 that the latter, is a better approximation to the slope of the tangent than is the former. The error associated with either finite difference analog depends on the shape of the curve at the point (x_i, u_i); but, in general, a second-order-correct analog is always a better approximation than is a first-order-correct analog.

The finite difference analog to the second derivative at (x_i, u_i) can also be interpreted geometrically. The slope of the chord between the points (x_{i+1}, u_{i+1}) and (x_i, u_i) is a second-order-correct analog of the slope of the curve at $(x_{i+1/2}, u_{i+1/2})$, the point on the curve halfway between the ends of this chord. Likewise, the slope of the chord between (x_i, u_i) and (x_{i-1}, u_{i-1}) is a second-order-correct analog to the slope of the curve at $(x_{i-1/2}, u_{i-1/2})$. The difference between these two slopes, divided by the increment between $x_{i+1/2}$ and $x_{i-1/2}$, should be a second-order-correct analog to the second derivative at the point x_i. It is, in fact, identical to the analog given by equation (1-7a).

9. FINITE DIFFERENCE EQUATIONS

If, now, the second-order-correct analogs to the first and second derivatives are substituted into equation (1-1) for the derivatives, the finite difference equation will be

$$\frac{u_{i+1} - 2u_i + u_{i-1}}{(\Delta x)^2} + \frac{2}{q + i(\Delta x)} \cdot \frac{u_{i+1} - u_{i-1}}{2(\Delta x)} - \frac{2p}{q + i(\Delta x)} u_i = 0 \quad (1\text{-}9)$$

When rearranged, it becomes

$$\left[1 - \frac{\Delta x}{q + i(\Delta x)}\right] u_{i-1} + \left[-2 - \frac{2p(\Delta x)^2}{q + i(\Delta x)}\right] u_i + \left[1 + \frac{\Delta x}{q + i(\Delta x)}\right] u_{i+1} = 0 \quad (1\text{-}9a)$$

This equation contains three unknowns: u_{i-1}, u_i, and u_{i+1}. (The values of the dependent variable at all points other than the boundary points are unknown.) An equation similar to (1-9a) can be written for each of the interior points. There are $R + 1$ points for R increments, and $R - 1$ equations can be written. The boundary conditions, equations (1-2), specify u_0 and u_R, the values of the dependent variable at the end points. Thus, the $R - 1$ unknowns corresponding to the values of the dependent variable at the interior points are determined by the $R - 1$ equations.

10. BOUNDARY CONDITION EQUATIONS

The finite difference equations written about the points x_1 and x_{R-1} are slightly different from equation (1-9a). In each case, the value of u is given at one point, and therefore it is not an unknown. These two relations are

$$\left[-2 - \frac{2p(\Delta x)^2}{q + \Delta x}\right] u_1 + \left(1 + \frac{\Delta x}{q + \Delta x}\right) u_2 = 0 \quad (1\text{-}10)$$

and

$$\left[1 - \frac{\Delta x}{q + (R-1)(\Delta x)}\right] u_{R-2} + \left[-2 - \frac{2p(\Delta x)^2}{q + (R-1)(\Delta x)}\right] u_{R-1} = -\left[1 + \frac{\Delta x}{q + (R-1)(\Delta x)}\right] \quad (1\text{-}11)$$

The set of $R - 1$ equations which determines the values of u at the $R - 1$ interior points comprises (1) equation (1-10), (2) $R - 3$ equations of the form of equation (1-9a) for which the index i assumes integral values between 2 and $R - 2$, and (3) equation (1-11). Equations of this form are readily solved by an algorithm of Thomas.

11. ALGORITHM OF THOMAS FOR TRIDIAGONAL MATRIX

The set of equations which has been obtained is of the form

$$\left.\begin{aligned}
&b_1u_1 + c_1u_2 + 0 + \cdots + 0 = d_1 \\
&a_2u_1 + b_2u_2 + c_2u_3 + 0 + \cdots + 0 = d_2 \\
&\qquad\qquad\cdots \\
&0 + \cdots + a_iu_{i-1} + b_iu_i + c_iu_{i+1} + \cdots + 0 = d_i \\
&\qquad\qquad\cdots \\
&0 + \cdots + 0 + a_{R-2}u_{R-3} + b_{R-2}u_{R-2} + c_{R-2}u_{R-1} = d_{R-2} \\
&0 + \cdots + 0 + a_{R-1}u_{R-2} + b_{R-1}u_{R-1} = d_{R-1}
\end{aligned}\right\} \tag{1-12}$$

A method of solving a set of equations of this type has been developed by Thomas. This method is very efficient for use on the digital computer and is presented in the Appendix. This method has been found to be stable to round-off error for finite difference equations of this type.

12. OTHER TYPES OF BOUNDARY CONDITIONS

The numerical technique described above can be readily applied to problems in which the first derivative is specified at one boundary of the region. As an example, consider a problem defined by equation (1-1) with the boundary conditions

$$\frac{du}{dx}(0) = g \tag{1-13}$$

$$u(1) = 1 \tag{1-2b}$$

These conditions arise when the rate of heat transferred to the rod is specified at one end. For R equal increments of x there are now R unknown values of u, since u_0 is not specified by the boundary condition of equation (1-13). There must therefore be R finite difference equations to specify the R unknowns. Equation (1-11) will still apply to this problem, since the boundary condition at $x = 1$ is unchanged. Also, there will be $R - 2$ equations of the form of equation (1-9a), for which the index i takes on values from 1 to $R - 2$. It remains to write one more equation about the point x_0, where $x = 0$.

The boundary condition specified here is that the first derivative, du/dx, is equal to the constant value g. To specify the value of the first derivative at the boundary, a fictitious point, x_{-1}, outside the region is used. The second-order-correct analog to $(du/dx)_0$, given by equation (1-8) with $i = 0$, is then set equal to g. Thus,

$$u_{-1} = u_1 - 2g(\Delta x) \tag{1-14}$$

The boundary equation is obtained by substituting this value for u_{-1} into equation (1-9a) for $i = 0$. The resulting equation is

$$\left[-2 - \frac{2p(\Delta x)^2}{q}\right]u_0 + 2u_1 = 2g(\Delta x)\left(1 - \frac{\Delta x}{q}\right) \tag{1-15}$$

Equations (1-1), (1-13), and (1-2b) which describe the continuous system can thus be replaced by equations (1-9a) for $1 \leq i \leq (R - 2)$, (1-15), and (1-11), which form a discrete system. These equations of the discrete system form a tridiagonal set and can be solved by the Thomas algorithm.

An even more general boundary condition results when the rate of heat transferred to an end of the rod is given by Newton's law of convection. In this case the boundary condition at $x = 0$ can be expressed as

$$\frac{du}{dx}(0) - Hu(0) = -g \tag{1-16}$$

where

$$H = \frac{hL}{k} \tag{1-16a}$$

The condition of equation (1-16) for $x = 0$ is handled similarly to that for equation (1-13). A fictitious point is used, and the dependent variable at this point, u_{-1}, is eliminated from the finite difference equation for $x = 0$ by the finite difference analog of the boundary condition. This is

$$\frac{u_1 - u_{-1}}{2(\Delta x)} - Hu_0 = -g \tag{1-17a}$$

or

$$u_{-1} = u_1 - 2H(\Delta x)u_0 + 2g(\Delta x) \tag{1-17b}$$

The resulting finite difference equation is

$$\left[-2 - 2H(\Delta x) - \frac{2(p - H)(\Delta x)^2}{q}\right]u_0 + 2u_1 = -2g(\Delta x)\left(1 - \frac{\Delta x}{q}\right) \tag{1-18}$$

The equations for this problem also are of the tridiagonal form and can be readily solved by the Thomas algorithm.

It should be observed at this point that the use of the fictitious point is merely a convenient way to obtain a correct analog at the boundary. The use of this point does not imply that the differential equation is applied to the region outside the boundary. Another analysis of equation (1-18) as the analog to equation (1-1) with the boundary condition of equation (1-16) demonstrates this point. For the analog to du/dx in equation (1-1) at the boundary $x = 0$, its equivalent from equation (1-16), $Hu_0 - g$, is used. A first-order-correct analog to d^2u/dx^2 at the boundary is used. This is

$$\left(\frac{d^2u}{dx^2}\right)_{x=0} = \frac{\dfrac{u_1 - u_0}{\Delta x} - \left(\dfrac{du}{dx}\right)_{x=0}}{\Delta x/2} \tag{1-16b}$$

The value of du/dx at $x = 0$ from equation (1-16) is again used in this analog. The finite difference equation which results from this treatment is equation (1-18). The use of the fictitious point is more convenient, though it does not demonstrate the order correctness of the equation. A similar analysis can be used in place of the fictitious point in all other cases.

When heat is transferred to one end of the rod by radiation, the resulting boundary condition is a nonlinear relationship between u and (du/dx) at the boundary. Treatment of nonlinear boundary conditions is covered in a later chapter.

13. POINTS SHIFTED FROM THE BOUNDARY

It can be noted that the number of unknown values of u, for a given number of increments, depends on the type of boundary conditions. For example, there are $R - 2$ unknowns with R increments for the boundary conditions of equation (1-2), and there are $R - 1$ for the conditions of (1-2b) with either (1-13) or (1-16). For many cases this variation is not inconvenient. However, it is sometimes desirable to develop a single set of equations which will handle a variety of boundary conditions. This objective can be accomplished by spacing the first point one-half an increment from the boundary. Such an arrangement of points also proves valuable in several other situations. In any event, it presents an alternate way to set up the finite difference equations.

The points are arranged as shown in Figure 1-4, and the value of the independent variable at each point is given by

$$x_i = (i - \tfrac{1}{2})\,\Delta x \tag{1-19}$$

In this arrangement there are no points on the boundaries; so the number of unknown values of u is equal to the number of increments for all types of boundary conditions. Also, there are two points outside the boundaries, one at each end. These are shown as x_0 and x_{R+1} in Figure 1-4.

Let us develop the finite difference equations for equation (1-1) with the boundary conditions of equations (1-2b) and (1-16). The finite difference

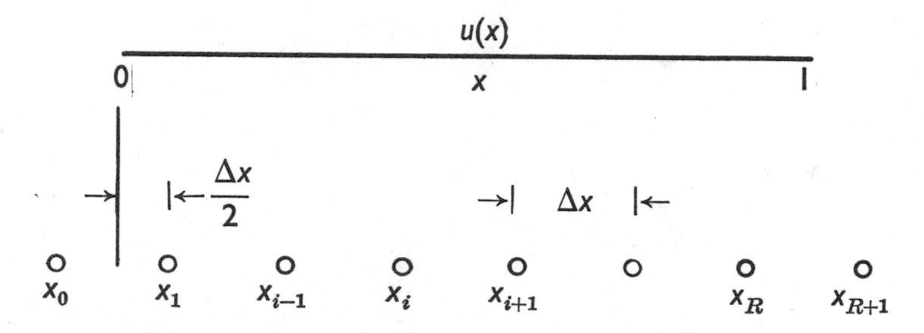

Figure 1-4. Grid points shifted from boundaries.

equation (1-9a) with i replaced by $(i - \frac{1}{2})$ in the denominators can be written for each of the points for $1 \leq i \leq R$. This equation is

$$\left[1 - \frac{\Delta x}{q + (i - \frac{1}{2})\,\Delta x}\right]u_{i-1} + \left[-2 - \frac{2p(\Delta x)^2}{q + (i - \frac{1}{2})\,\Delta x}\right]u_i + \left[1 + \frac{\Delta x}{q + (i - \frac{1}{2})\,\Delta x}\right]u_{i+1} = 0 \tag{1-9b}$$

The values of u at the two exterior points, u_0 and u_{R+1}, must be eliminated from these equations by the boundary conditions. The value of u at $x = 1$ is specified, but there is no grid point at this value of x for the arrangement of Figure 1-4. A second-order-correct analog to this value is the average of the values of u on either side of the boundary. Thus, the analog to this boundary condition is

$$\frac{u_R + u_{R+1}}{2} = 1 \tag{1-20}$$

An expression for u_{R+1} is obtained from this relation, and the result is substituted into equation (1-9b) for $i = R$. The resulting equation becomes

$$\left[1 - \frac{\Delta x}{q + (R - \frac{1}{2})\,\Delta x}\right]u_{R-1} + \left[-3 - \frac{2p(\Delta x)^2 + \Delta x}{q + (R - \frac{1}{2})\,\Delta x}\right]u_R = -2\left[1 + \frac{\Delta x}{q + (R - \frac{1}{2})\,\Delta x}\right] \quad (1\text{-}21)$$

The boundary condition at the other end is handled in a similar manner. The value of u at $x = 0$ is approximated as the average of u_0 and u_1. The derivative at this point is approximated as

$$\left(\frac{du}{dx}\right)_{1/2} \approx \frac{u_1 - u_0}{\Delta x} \quad (1\text{-}22)$$

This analog is second-order correct, since it is used to approximate the derivative at the center of the interval, although it is written over only a single interval. A geometrical interpretation of this analog is given in Section 8 in the discussion of the analog to the second derivative. This type of analog to the first derivative at the center of an interval is used extensively in the numerical solution of first-order partial differential equations. These equations are discussed in Chapter Three.

The finite difference analog to the boundary condition of equation (1-16) at $x = 0$ then becomes

$$\frac{u_1 - u_0}{\Delta x} - H\frac{u_1 + u_0}{2} = -g \quad (1\text{-}23a)$$

Written explicitly for u_0, it becomes

$$u_0 = \left[\frac{2 - H(\Delta x)}{2 + H(\Delta x)}\right]u_1 + \frac{2g(\Delta x)}{2 + H(\Delta x)} \quad (1\text{-}23b)$$

The resulting finite difference equation is

$$\left[-\frac{\Delta x}{q + \frac{1}{2}\Delta x}\left(\frac{2 - H\,\Delta x}{2 + H\,\Delta x} + 2p\,\Delta x\right) - \frac{2 + 3H\,\Delta x}{2 + H\,\Delta x}\right]u_1 + \left(1 + \frac{\Delta x}{q + \frac{1}{2}\Delta x}\right)u_2 = -\left(1 + \frac{\Delta x}{q + \frac{1}{2}\Delta x}\right)\left(\frac{2g\,\Delta x}{2 + H\,\Delta x}\right) \quad (1\text{-}24)$$

The complete set of finite difference equations for the differential equation (1-1) with the boundary conditions of equations (1-16) and (1-2b) for the arrangement of points of Figure 1-4 is made up of (1) equation (1-24), (2)

$R - 2$ equations of the form of equation (1-9b) for $2 \leq i \leq (R - 1)$, and (3) equation (1-21). These equations are of the form of equations (1-12) and can be solved by the Thomas algorithm.

Further discussion of these boundary conditions will bring out another advantage of the spacing of points as shown in Figure 1-4. The parameter H is directly proportional to the coefficient of heat transfer between the end of the rod and its surroundings, as shown in its definition in equation (1-16a). By proper adjustment of this parameter, one can simulate, both mathematically and physically, the boundary conditions of equation (1-2a) and of equation (1-13) by the more general condition of equation (1-16). When H is set to zero, the physical boundary condition specifies that the rate of convection is zero, and the heat flux may be specified. For $H = 0$, the mathematical boundary condition, equation (1-16), becomes identical with equation (1-13), the mathematical boundary condition which specifies the flux. Similarly, when H and g are made very large, the physical condition is a specification of the temperature at $x = 0$. For these values of H and g, equation (1-16) becomes identical with equation (1-2a); the ratio of g to H determines the value of u which is specified. For the arrangement of grid points shown in Figure 1-4, the finite difference analogs to the boundary conditions behave similarly. Thus, equation (1-23b) becomes a specification of du/dx when $H = 0$ and a specification of u when H and g are large. This same generalization of the finite difference analogs to the boundary condition does not occur so readily when the points are arranged as shown in Figure 1-2.

One disadvantage of the arrangement of points of Figure 1-4 is that the temperature at the boundaries is not computed. However, each of these can be evaluated as the average of the exterior value and the last interior value. The exterior value can be obtained from the appropriate boundary condition analog.

14. EQUATION IN RADIAL COORDINATES

The arrangement of points shown in Figure 1-4 is also convenient to use when the problem is described by radial coordinates. The differential equation for heat conduction in radial coordinates is similar to the equation for a tapered rod which goes to a point at one end. This latter equation is obtained by setting the parameter q in equation (1-1) to zero. In fact, when the parameter p is also set to zero so that there is no heat loss from the surface of the rod by convection, equation (1-1) becomes the equation for heat conduction in a sphere. The equation for heat conduction in a cylinder is similar. Although numerical methods can be used to great advantage for unsteady-state heat conduction in a cylinder or sphere, the steady-state problems for these geometries have trivial solutions. Therefore, to study the treatment of radial coordinates on a steady-state problem, let us consider a tapered rod which goes to a point at one end.

The governing differential equation is

$$\frac{d^2u}{dx^2} + \frac{2}{x}\left(\frac{du}{dx} - pu\right) = 0 \tag{1-1a}$$

The boundary condition at $x = 1$ depends on the physical condition imposed. Let us use the simple one of equation (1-2b), which is

$$u(1) = 1 \tag{1-2b}$$

The boundary condition at $x = 0$ is determined by the nature of equation (1-1a). For the second term of this equation to remain finite when x goes to zero, the numerator of this term—namely, $(du/dx) - pu$—must also go to zero. Thus, the boundary condition at $x = 0$ is

$$\frac{du}{dx}(0) - pu(0) = 0 \tag{1-16c}$$

which is identical to equation (1-16) with $H = p$ and $g = 0$. The finite difference analogs to the equations (1-1a), (1-2b), and (1-16c) can be obtained directly from equations (1-9b), (1-21), and (1-24) by setting $q = 0$, $g = 0$, and $H = p$. Thus, no complication arises at $x = 0$ for radial coordinates when the points are arranged as shown in Figure 1-4.

The conversion is not so simple when the points are arranged as shown in Figure 1-2. Equation (1-9a) is still a satisfactory analog to equation (1-1a) for all points that are completely surrounded by interior points, and equation (1-11) can still be used for $i = R - 1$. However, the equation at the other boundary, for $i = 0$, cannot be directly obtained from equation (1-18). Two terms of this equation become indeterminant when the proper values of the parameters are introduced. This complication actually arises when the differential equation (1-1a) is applied at $x = 0$, for the second term becomes indeterminant in this case. The correct differential equation to be applied at $x = 0$ is obtained by applying L'Hôpital's rule to the second term of (1-1a). The resulting equation is

$$3\frac{d^2u}{dx^2} - 2p\frac{du}{dx} = 0 \tag{1-1b}$$

The finite difference analog to this equation is obtained by substituting the analogs of equations (1-7a) and (1-8) with $i = 0$ into equation (1-1b). The value of u_{-1} from (1-17b) with $H = p$ and $g = 0$ is then substituted into this equation to obtain the final equation for $i = 0$. The resulting equation is

$$[-6 - 6p(\Delta x) - 2p^2(\Delta x)^2]u_0 + 6u_1 = 0 \tag{1-25}$$

The equations which complete the set are

$$(i - 1)u_{i-1} + [-2i - 2p(\Delta x)]u_i + (i + 1)u_{i+1} = 0 \tag{1-9c}$$

for $1 \leq i \leq (R - 2)$, and

$$(R - 2)u_{R-2} + [-2(R - 1) - 2p(\Delta x)]u_{R-1} = -R \tag{1-11a}$$

The radial problem can thus also be handled with grid points on the boundaries by use of the proper equations. The several methods developed in this chapter for steady-state problems will be applied to unsteady-state problems in the next chapters.

Chapter Two

LINEAR PARABOLIC PARTIAL DIFFERENTIAL EQUATIONS

1. DIFFERENTIAL EQUATION CONSIDERED

Parabolic partial differential equations arise from unsteady-state problems in which transport by conduction or diffusion is important. A general equation of this type describes the unsteady-state heating of the tapered rod considered in Chapter One. This equation can be obtained from equation (1-1) by converting the derivatives with respect to length to partial derivatives and inserting the partial derivative with respect to time on the right side of that equation. The methods for numerical solution of partial differential equations could be demonstrated on the resulting equation. However, the new ideas which arise in the treatment of partial differential equations can be most readily demonstrated on the simplest parabolic equation. This equation describes conduction in a uniform, insulated rod; it is

$$\frac{\partial^2 u}{\partial x^2} = \frac{\partial u}{\partial t} \tag{2-1}$$

The treatment of various types of boundary conditions has also been presented in Chapter One; so let us consider only the simplest boundary conditions for purposes of demonstrating the techniques for partial differential equations. These boundary conditions are

$$\left.\begin{aligned} u(0, t) &= 0 \\ u(1, t) &= 1 \end{aligned}\right\} \quad \text{all } t \tag{2-2}$$

Conditions which define the dependent variable at the initial time are also required for definition of the problem. The initial conditions to be used are

$$u(x, 0) = 0; \qquad x < 1 \tag{2-3}$$

2. NOMENCLATURE FOR FUNCTION OF TWO DISCRETE VARIABLES

The region described by two independent, continuous variables is a part of a plane. For the problem under consideration, the length variable, x, varies between 0 and 1, and the time variable, t, increases without limit from zero. Thus, the portion of the (x, t) plane described is the semi-infinite strip in the positive half-plane between the lines $x = 0$ and $x = 1$. This is the shaded area in Figure 2.1. When these two independent variables are replaced by discrete variables (also called x and t), the new variables are defined at points located as shown in Figure 2-1. The region between 0 and 1 along the x-axis is divided into R equal increments of size Δx, with grid points being placed on each boundary. The points could also be spaced so that one point would be $\Delta x/2$ from each boundary as in the arrangement of Figure 1-4. However, the arrangement shown in Figure 2-1 will be used for the development given

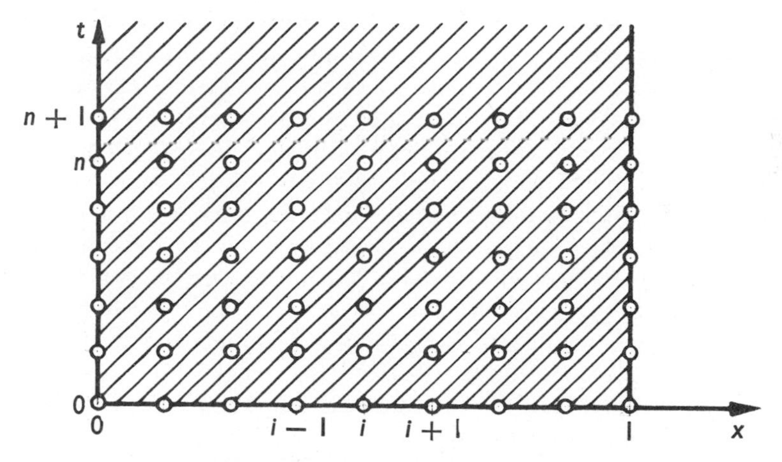

Figure 2-1. Grid points for unsteady-state problem.

in this chapter. The time axis is divided into increments of size Δt. For many numerical solutions, it will be desirable to increase the size of the time step, Δt, as the solution progresses. Consequently, the discrete points may be unevenly spaced along the t-axis.

The index i is used to indicate position along the x-axis, and the index n is used for the t-axis. The value of the discrete variable x at a given point is given by equation (1-5), and equations (1-6) hold for adjacent values of x. Since the values of the time increment, Δt, may not be constant, the value of time is given by

$$t_n = \sum_{m=1}^{n} (\Delta t)_m \tag{2-4}$$

The dependent variable, u, is now a function of two independent variables,

x and t. It is therefore necessary to use two subscripts to specify the value of u at a given point; thus $u(x_i, t_n) = u_{i,n}$.

3. GENERAL PROCEDURE FOR SOLVING PARABOLIC EQUATIONS

For the simple problem discussed in Chapter One, there was a single row of points at which the dependent variable was unknown. These values were found by solving simultaneously the finite difference equations written for each of these points. For the problem described by equations (2-1), (2-2), and (2-3), the value of the dependent variable is unknown at a row of points at each time level, and there are actually an unlimited number of time levels. It is not feasible to solve for all the unknown values of u simultaneously even when a limited number of time levels are considered. Consequently, the technique employed is to solve for the unknown values of u at one time level, using the known values of u at the previous time level. The values of u at the initial time level, where $n = 0$, are given by the initial conditions of equation (2-3). These values are used to determine the unknown values of u at the next time level for which $n = 1$. The same computational procedure is then used to find the values of u for $n = 2$ from the now known values of u at $n = 1$. This procedure is continued for as many time increments as desired. Therefore, the finite difference equations are formulated so that they contain values of u at two consecutive time levels. The index n is used to designate the last time level at which the values of u are known, and the index $n + 1$ is used to designate the next time level at which values of the dependent variable are unknown. In following finite difference equations, therefore, any dependent variable subscripted with $n + 1$ will be an unknown value unless it is given by a boundary condition. Likewise, any dependent variable subscripted with n will be a known value.

4. FORWARD DIFFERENCE EQUATION

A number of finite difference equations analogous to equation (2-1) can be written, and it might be instructive to examine several of these. The forward or explicit difference equation is probably the most well known, although it is the least efficient of all the possible equations which can be used. However, it is of value to examine this equation as the first logical step in developing the more efficient methods.

In developing the forward finite difference equation, one writes the analog of $\partial^2 u/\partial x^2$ developed in the preceding chapter at the known time level which is indexed by n. This relation is

$$\left(\frac{\partial^2 u}{\partial x^2}\right)_{i,n} \approx \frac{u_{i+1,n} - 2u_{i,n} + u_{i-1,n}}{(\Delta x)^2} \tag{2-5}$$

This analog is second-order correct in the variable x. The analog to the time derivative is obtained from a Taylor series in time about the point x_i, t_n; this is

$$\left(\frac{\partial u}{\partial t}\right)_{i,n} = \frac{u_{i,n+1} - u_{i,n}}{\Delta t} - \left(\frac{\partial^2 u}{\partial t^2}\right)_{i,n} \frac{\Delta t}{2!} \cdots \tag{2-6}$$

The analog resulting from the truncation of this series after the first term is first-order correct in t.

When the analogs defined in equations (2-5) and (2-6) are substituted into equation (2-1), the finite difference equation is

$$u_{i,n+1} = \frac{\Delta t}{(\Delta x)^2}(u_{i+1,n} + u_{i-1,n}) + \left[1 - \frac{2\,\Delta t}{(\Delta x)^2}\right]u_{i,n} \tag{2-7}$$

This equation contains only one unknown value, $u_{i,n+1}$, and is written explicitly for this unknown. The numerical computation of the values of the dependent variable is thus made one point at a time.

Equation (2-7) is certainly a simple one to formulate, and it is especially easy to use for computing the values of u at each time level. However, for a numerical solution to be of any value, its solution must converge to the solution of the corresponding differential equation when the finite increments, Δx and Δt, are decreased in size. Analysis has shown that a very restrictive relationship between the size of Δx and that of Δt must be satisfied in order for the solution of equation (2-7) to approach that of equation (2-1). This analysis is in a later section of this chapter for all three of the analogs to equation (2-1) which are developed. At this point, let us examine the consequences of this restriction and look at some physical interpretations of it.

The restriction requires that the ratio of Δt to $(\Delta x)^2$ must remain less than or equal to $\frac{1}{2}$. This restriction is a rather serious one, for, in order to minimize the truncation error in the x analogs, the size of Δx must be small. The many transient problems which have boundary conditions independent of time approach a steady-state condition. For these problems, including the one defined by equations (2-1), (2-2), and (2-3), the time increment can be continuously increased and made quite large as the solution progresses toward steady state without causing significant truncation error in the time analog. However, for the forward difference equation, the size of Δt must remain on the order of $(\Delta x)^2$ for the solution to be stable. Thus, a very small value of Δt must be used for stability even when a much larger value could be used without causing truncation error. A finite difference equation which does not have this restriction is therefore a much better one to use as an analog to equation (2-1).

Before obtaining this equation, however, let us examine this restriction from several viewpoints. The restriction requires that the coefficient of $u_{i,n}$ in equation (2-7) remain positive or, at least, zero. It is certainly reasonable for this restriction to hold; for, if this coefficient were negative, the values of u_i might well oscillate from one time step to the next. This behavior is, in fact, observed when the restriction is violated. However, it must be pointed out that the condition of a positive coefficient on $u_{i,n}$ cannot be used as a criterion for stability. It is merely a reasonable consequence of the restriction established by stability analysis.

A comparison of the solution of equation (2-7) with the physical problem described by equation (2-1) is also helpful in understanding the restriction for stability. From the physical problem, it is known that the temperature near the cold end of the bar will begin to rise from its initial temperature rather quickly after the other end is heated. Just how fast this temperature rises will depend on the values of the parameters contained in the dimensionless time variable. However, a sensitive thermocouple would indicate a slight temperature rise near the cold end very soon after the other end was heated.

Let us examine the solution of equation (2-7) with the initial and boundary conditions of the problem. For the initial time level ($n = 0$), $u_{i,0} = 0$ for $0 \leq i \leq R - 1$, and at the hot end $u_{R,0} = 1$. Therefore when equation (2-7) is solved for u at the first unknown time level, $u_{i,1} = 0$ for all interior points except $u_{R-1,1}$. For the next time level, only $u_{R-1,2}$ and $u_{R-2,2}$ will be other than zero. At each succeeding time level, only one additional value of temperature will rise above its initial value of zero. If the bar is divided into R increments, a total of $R - 1$ time intervals must be computed before the temperature at the next to the last point, u_1, will rise above its initial value. Very small time increments must therefore be used so that values of u computed from equation (2-7) will be representative of the physical problem.

5. BACKWARD DIFFERENCE EQUATION

In searching for a new finite difference equation which does not have a restriction on the size of Δt for stability, we might write the finite difference analogs for $\partial^2 u/\partial x^2$ at the new or unknown time level which is indexed by $n + 1$. This backward difference is

$$\left(\frac{\partial^2 u}{\partial x^2}\right)_{i,n+1} = \frac{u_{i+1,n+1} - 2u_{i,n+1} + u_{i-1,n+1}}{(\Delta x)^2} \tag{2-8}$$

The time derivative analog is obtained from a Taylor series in time about the point x_i, t_{n+1}. This series is

$$\left(\frac{\partial u}{\partial t}\right)_{i,n+1} = \frac{u_{i,n+1} - u_{i,n}}{\Delta t} + \left(\frac{\partial^2 u}{\partial t^2}\right)_{i,n+1} \frac{\Delta t}{2} - \cdots - \tag{2-9}$$

The first term on the right side of this equation is a first-order-correct analog to the time derivative. When these two analogs are substituted into equation (2-1), the finite difference equation is

$$u_{i-1,n+1} + \left[-2 - \frac{(\Delta x)^2}{\Delta t}\right] u_{i,n+1} + u_{i+1,n+1} = -\frac{(\Delta x)^2}{\Delta t} u_{i,n} \tag{2-10}$$

This equation is implicit; that is, it contains three values of the dependent variable, u, at the unknown time level. These three unknown values are arranged in exactly the same order as those in equation (1-9a).

For R increments in x and the boundary conditions of equation (2-2) there will be $R - 1$ unknown values of u at each time level, and there are $R - 1$ equations, one corresponding to each of these points. The $R - 3$ equations for $2 \leq i \leq R - 2$ are of the form of equation (2-10). The equations for $i = 1$ and $i = R - 1$ for the boundary conditions of equation (2-2) are

$$\left[-2 - \frac{(\Delta x)^2}{\Delta t}\right] u_{1,n+1} + u_{2,n+1} = \left[-\frac{(\Delta x)^2}{\Delta t}\right] u_{1,n} \tag{2-11}$$

$$u_{R-2,n+1} + \left[-2 - \frac{(\Delta x)^2}{\Delta t}\right] u_{R-1,n+1} = \left[-\frac{(\Delta x)^2}{\Delta t}\right] u_{R-1,n} - 1 \tag{2-12}$$

The resulting set of equations is of the form of equations (1-12); that is, the coefficient matrix is tridiagonal. The equations can thus be readily solved by the Thomas method to obtain the $R - 1$ values of $u_{i,n+1}$. The same method can then be applied to obtain the values of $u_{i,n+2}$ from the now known values of $u_{i,n+1}$.

Analysis has shown that there is no restriction on the size of Δt for stability of the backward difference equation. The size of Δt can therefore be set, independently of the size of Δx, to minimize truncation error of the Taylor series in time. An examination of the backward difference equation will show that all the values of u at the first unknown time level ($n = 1$) will be changed from the initial values ($n = 0$). The boundary condition at $x = 1$ affects the values of the dependent variable at all positions for the new time step through the simultaneous solution of the difference equations by the Thomas method.

This behavior is characteristic of the physical problem and of the solution of the governing differential equation.

In the solution of these equations on a digitial computer, it is usual practice to keep the size of Δx constant throughout the calculation. The same program can then be used for computing the values of the dependent variable at each time level. If the boundary conditions are of the form of equation (2-2) so that the transient solution approaches a steady-state condition, it is usual practice to increase the size of Δt as the solution progresses. The solution is thus obtained in a minimum of computing time. Such practice can be justified by a simple analysis. As the steady-state conditions are approached, the difference in values of u from one time level to the next diminishes if a constant time increment is used. An examination of the Taylor series in equation (2-9) leads to the same conclusion. As the steady-state conditions are approached, the size of $(\partial^2 u/\partial t^2)$ and the higher time derivatives decrease; thus, the same truncation error will result from a larger Δt as the steady state is approached.

The backward difference is an efficient one, and it is simple to use. However, it is only first-order correct in time. It is desirable to find a second-order correct analog to this derivative.

6. CRANK-NICOLSON EQUATION

One method for obtaining a second-order-correct analog for $\partial u/\partial t$ would be to use a finite difference over two increments. Since the finite differences on x are written about the point x_i, t_{n+1}, the analog for $\partial u/\partial t$ should also be written about this point. The resulting second-order-correct analog is

$$\left(\frac{\partial u}{\partial t}\right)_{i,n+1} = \frac{u_{i,n+2} - u_{i,n}}{2(\Delta t)} - \left(\frac{\partial^3 u}{\partial t^3}\right)_{i,n+1} \frac{(\Delta t)^2}{6} - \cdots - \qquad (2\text{-}13)$$

This analog introduces values of the dependent variable at another unknown time level, that where the time index has the value $n + 2$. The method of calculating values of u at one unknown time level and then proceeding to the next therefore cannot be used. However, this analog has a much more serious drawback, for Douglas (*1*) has shown this analog to be unstable for any ratio of Δx to Δt. Another second-order-correct analog must thus be found.

The desired second-order-correct equation is called the Crank-Nicolson equation. For this equation all the finite differences are written about the point x_i, $t_{n+1/2}$, which is halfway between the known and the unknown time levels. In Figure 2-2, this point is shown as a cross. Values of the dependent variable, u, are computed only at the points designated by circles. The second-order-correct analog of the time derivative at the point x_i, $t_{n+1/2}$ is

$$\left(\frac{\partial u}{\partial t}\right)_{i,n+1/2} = \frac{u_{i,n+1} - u_{i,n}}{\Delta t} - \left(\frac{\partial^3 u}{\partial t^3}\right)_{i,n+1/2} \frac{(\Delta t)^2}{24} - \cdots - \tag{2-14}$$

This analog is actually the same as the first-order-correct analog used for both the backward difference equation and the forward difference equation. The analog is now second-order-correct, since it is used to approximate the derivative at the point x_i, $t_{n+1/2}$. A geometric interpretation can be obtained by referring to Figure 1-3.

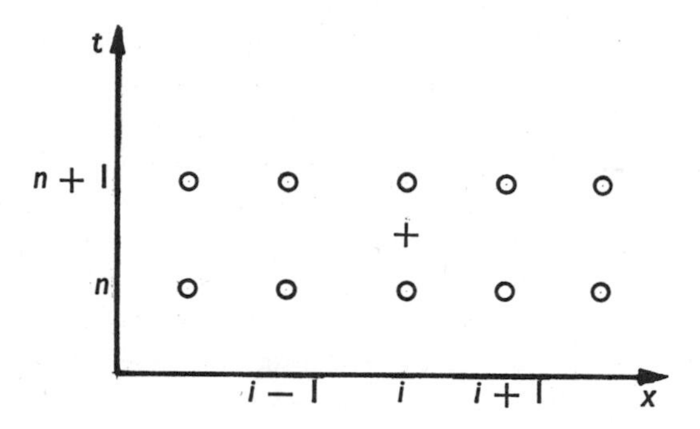

Figure 2-2. Center of analogs for Crank-Nicolson equation.

The real key to the Crank-Nicolson equation is the manner of approximating $\partial^2 u/\partial x^2$ without requiring the evaluation of the dependent variable at the time level for which the time index has the value $n + 1/2$. This derivative is approximated by the arithmetic average of its finite difference analogs at the points x_i, t_n, and x_i, t_{n+1}. The resulting analog is the average of the forward and backward analogs and is

$$\left(\frac{\partial^2 u}{\partial x^2}\right)_{i,n+1/2} \approx \frac{1}{2}\left[\frac{u_{i+1,n} - 2u_{i,n} + u_{i-1,n}}{(\Delta x)^2} + \frac{u_{i+1,n+1} - 2u_{i,n+1} + u_{i-1,n+1}}{(\Delta x)^2}\right] \tag{2-15}$$

The Crank-Nicolson finite difference equation obtained from substituting the analogs of equations (2-14) and (2-15) into equation (2-1) is second-order correct in both x and t. It is

$$u_{i-1,n+1} + \left[-2 - \frac{2(\Delta x)^2}{\Delta t}\right]u_{i,n+1} + u_{i+1,n+1}$$
$$= -u_{i-1,n} + \left[2 - \frac{2(\Delta x)^2}{\Delta t}\right]u_{i,n} - u_{i+1,n} \quad (2\text{-}16)$$

The boundary equations can be obtained from equation (2-16) by setting $u_{0,n+1} = u_{0,n} = 0$ in the equation for $i = 1$, and $u_{R,n+1} = u_{R,n} = 1$ in the equation for $i = R - 1$. These boundary conditions correspond to those given by equation (2-2). Equation (2-16) contains the same three unknown values of u that are found in the backward difference equation. This equation and its two boundry equations are also of the form of equations (1-12) of the previous chapter; thus, they can readily be solved by the Thomas method. The Crank-Nicolson method requires more computation per time step than the backward difference method to evaluate the right side of the equations. The right side of equation (2-16) contains values of u at three length positions at the known time level instead of the one known value found in equation (2-10). However, a larger time increment can be used for the Crank-Nicolson equation, since its time derivative analog is second-order correct. Fewer time steps are thus necessary to compute values of the dependent variable for a given elapsed time. Thus, the Crank-Nicolson equation is more efficient than the backward difference method and is the preferred method for obtaining numerical solutions to parabolic differential equations.

The Crank-Nicolson equation, like the backward difference equation, is stable for all ratios of Δx to Δt. As a consequence, it can be shown that a stable solution can be obtained even when an unstable equation is used for half of the time steps. The Crank-Nicolson equation is the result of successive applications of the forward and backward equations. This fact can be demonstrated by writing the backward equation between the $n + 1$ and $n + 2$ time levels. The result is

$$u_{i-1,n+2} + \left[-2 - \frac{(\Delta x)^2}{\Delta t}\right]u_{i,n+2} + u_{i+1,n+2} = -\frac{(\Delta x)^2}{\Delta t}u_{i,n+1} \quad (2\text{-}10a)$$

The value of $u_{i,n+1}$ from the forward equation, equation (2-7), is then substituted into (2-10a). The result is obviously the Crank-Nicolson equation written between the n and $n + 2$ time levels with an increment of $2(\Delta t)$. Therefore, every time the Crank-Nicolson equation is used, in effect, an unstable forward equation followed by a backward equation with the same time increment is being used. This factorization of a stable equation into two equations, one of which is unstable by itself, is used extensively in the solution of parabolic equations in two and three space dimensions. An understanding of this manipulation at this point will prepare the reader to understand more readily the methods for numerical solution of multidimensional problems to be presented in later chapters.

7. OTHER BOUNDARY CONDITIONS

The backward difference can be used satisfactorily with all types of boundary conditions, including those of equations (1-13) and (1-16). The boundary condition need be applied only at the unknown time level $(n + 1)$, since the value of the dependent variable at the fictitious point occurs in the finite difference equations only at this time level. The value at the fictitious point appears in the Crank-Nicolson equation at both time levels, however. In this case, the boundary conditions are applied separately at each time level (n and $n + 1$) to eliminate the values at the fictitious point. The Crank-Nicolson equation is also satisfactory with the boundary condition of equation (1-13) when the analog of equation (1-14) is used. However, when the Crank-Nicolson is used for the boundary condition of equation (1-16), with a grid point placed on the boundary, the computed value of u at the boundary oscillates for large values of H. This oscillation is stable about the correct value, and the other interior points are unaffected. Douglas (*1*) has analyzed this oscillation. The oscillation does not occur when the first interior point is located one-half increment from the boundary. If a point is located on the boundary, the backward difference equation must be used at this point. Unfortunately, for conditions which cause this oscillation, the time derivatives are large, so compensatingly smaller time increments must be used with the backward difference equation. Of course, if H is large enough, the boundary condition of equation (1-2) can be used.

8. STABILITY ANALYSIS

At this point, let us perform analyses of the stability of the three finite difference equations we have developed for the solution of equation (2-1). Errors are introduced by the truncation of the series which are used to represent the derivatives in the process of replacing the differential equation by finite difference equations. Let us analyze the growth of these errors and find the conditions for which the errors will be attenuated from one time step to the next. First, let us rewrite the differential equation under consideration. It is

$$\frac{\partial^2 u}{\partial x^2} = \frac{\partial u}{\partial t} \tag{2-1}$$

Now, let us replace these derivatives by the complete series which have been developed to represent them. When this is done with the series for the forward analogs, equation (2-1) becomes

$$\frac{u_{i+1,n} - 2u_{i,n} + u_{i-1,n}}{(\Delta x)^2} - \left(\frac{\partial^4 u}{\partial x^4}\right)_{i,n} \frac{2(\Delta x)^2}{4!} - \cdots - \\ = \frac{u_{i,n+1} - u_{i,n}}{\Delta t} - \left(\frac{\partial^2 u}{\partial t^2}\right)_{i,n} \frac{\Delta t}{2!} - \cdots - \tag{2-1a}$$

Equation (2-1a) is still a differential equation. For convenience, we shall denote the dependent variable by w in the finite difference equation used in place of equation (2-1a). This differential equation is to be represented by the forward finite difference equation:

$$\frac{w_{i+1,n} - 2w_{i,n} + w_{i-1,n}}{(\Delta x)^2} = \frac{w_{i,n+1} - w_{i,n}}{\Delta t} \tag{2-7a}$$

The error resulting from using equation (2-7a) in place of equation (2-1a) satisfies the following relation:

$$\frac{z_{i+1,n} - 2z_{i,n} + z_{i-1,n}}{(\Delta x)^2} - \left(\frac{\partial^4 u}{\partial x^4}\right)_{i,n} \frac{2(\Delta x)^2}{4!} - \cdots - \\ = \frac{z_{i,n+1} - z_{i,n}}{\Delta t} - \left(\frac{\partial^2 u}{\partial t^2}\right)_{i,n} \frac{\Delta t}{2!} - \cdots - \tag{2-17}$$

In this relation, the error, $z_{i,n}$ equals $u_{i,n} - w_{i,n}$. As a result, when the increments Δt and Δx are small, the error, $z_{i,n}$, satisfies the same finite difference equation as that used to represent the differential equation. The boundary conditions for the differential equation can often be applied exactly to the finite difference equation so that there is no error in the boundary conditions. Therefore, the boundary conditions for equation (2-17) are

$$z_{0,n} = z_{R,n} = 0 \tag{2-18}$$

It now remains to find a solution to the partial finite difference equation (2-17). The analytical solution to the differential equation (2-1) was obtained by separation of variables, and the same technique can be used on equation (2-17). Therefore, let

$$z_{i,n} = \rho_n \varphi_i \tag{2-19}$$

The variables can be separated, and the equation put in the form

$$\frac{\rho_{n+1}}{\rho_n} = \frac{\Delta t}{\varphi_i} \Delta_x^2 \varphi_i + 1 = \mu \tag{2-20}$$

In this equation μ is a constant, and

$$\Delta_x^2 \varphi_i = \frac{\varphi_{i+1} - 2\varphi_i + \varphi_{i-1}}{(\Delta x)^2}$$

It is necessary to analyze the behavior of the ratio ρ_{n+1}/ρ_n. This ratio is a measure of the growth of the error from one time level to the next. The absolute value of this ratio must be equal to or less than 1 in order for the solution to the finite difference equation (2-7a) to be stable.

The value of this ratio can be found from solutions of the ordinary difference equations

$$\Delta_x^2 \varphi_i + \frac{1-\mu}{\Delta t} \varphi_i = 0; \qquad i = 1, 2, \cdots, R-1 \tag{2-21}$$

with $\varphi_0 = 0$ and $\varphi_R = 0$. By analogy to the corresponding differential equation, the eigenfunctions for this set of equations are the sine and cosine functions. However, the coefficients for the cosines must be zero to satisfy the boundary conditions. Therefore,

$$\varphi_i^{(p)} = \sin \pi p x_i; \qquad p = 1, 2, \cdots, R-1 \tag{2-22}$$

It can be shown that

$$\Delta_x^2 \sin \pi p x_i = -\frac{4}{(\Delta x)^2} \sin^2 \left(\frac{\pi p \,\Delta x}{2}\right) \sin \pi p x_i$$

Thus, equation (2-21) becomes

$$\left[-\frac{4}{(\Delta x)^2} \sin^2 \left(\frac{\pi p \,\Delta x}{2}\right) + \frac{1-\mu}{\Delta t}\right] \sin \pi p x_i = 0$$

and, consequently, the eigenvalues are

$$\mu_p = 1 - \frac{4(\Delta t)}{(\Delta x)^2} \sin^2 \left(\frac{\pi p \,\Delta x}{2}\right); \qquad p = 1, 2, \cdots, R-1$$

The ratio ρ_{n+1}/ρ_n is equal to the eigenvalues, and its absolute value must not exceed unity. The term $\sin^2 (\pi p \,\Delta x/2)$ has a maximum value of 1, and it is at this value that the most stringent restriction on the ratio $\Delta t/(\Delta x)^2$ is obtained. Thus, for

$$\left[\frac{4\,\Delta t}{(\Delta x)^2} - 1\right] \leq 1$$

it is necessary that

$$\frac{\Delta t}{(\Delta x)^2} \leq \frac{1}{2}$$

This is exactly the restriction that was discussed in Section 4.

A similar analysis of the backward and Crank-Nicolson equations will show that there is no restriction on the increment ratio for stability for either equation. Consider first the backward equation. Again, the error must satisfy the same finite difference equation with no error at the boundaries, or

$$\frac{z_{i,n+1} - z_{i,n}}{\Delta t} = \Delta_x^2 z_{i,n+1}; \qquad i = 1, 2, \cdots, R-1$$

$$z_{0,n} = z_{R,n} = 0$$

The substitution of equation (2-19) is again used to separate the variables, with the result that

$$\frac{\rho_{n+1}}{\rho_n} = \frac{1}{1 - \dfrac{\Delta t}{\varphi_i} \Delta_x^2 \varphi_i} = \mu$$

and

$$\Delta_x^2 \varphi_i + \frac{1-\mu}{\mu\, \Delta t} \varphi_i = 0 \tag{2-23}$$

Equation (2-23) is the same as equation (2-21) with a different coefficient of φ_i, so the eigenfunctions of equation (2-22) also satisfy equation (2-23). The eigenvalues for this equation, however, are

$$\mu_p = \frac{1}{1 + \dfrac{4\,\Delta t}{(\Delta x)^2} \sin^2\left(\dfrac{\pi p\, \Delta x}{2}\right)}$$

From this relation it can be seen that ρ_{n+1}/ρ_n will remain less than 1 for all values of $\Delta t/(\Delta x)^2$.

Similarly, for the Crank-Nicolson equation the error must satisfy

$$\frac{z_{i,n+1} - z_{i,n}}{\Delta t} = \frac{1}{2} \Delta_x^2 (z_{i,n+1} + z_{i,n}); \qquad i = 1, 2, \cdots, R-1; \; z_{0,n} = z_{R,n} = 0$$

Again, the substitution of equation (2-19) separates the variables to yield

$$\frac{\rho_{n+1}}{\rho_n} = -\frac{\Delta_x^2 \varphi_i + \dfrac{2}{\Delta t} \varphi_i}{\Delta_x^2 \varphi_i - \dfrac{2}{\Delta t} \varphi_i} = \mu$$

and

$$\Delta_x^2 \varphi_i + \frac{1-\mu}{1+\mu} \frac{2}{\Delta t} \varphi_i = 0 \tag{2-24}$$

The eigenfunctions of equation (2-24) are also given by equation (2-22), and the eigenvalues are

$$\mu_p = \frac{1 - 2\dfrac{\Delta t}{(\Delta x)^2}\sin^2\left(\dfrac{\pi p\,\Delta x}{2}\right)}{1 + 2\dfrac{\Delta t}{(\Delta x)^2}\sin^2\left(\dfrac{\pi p\,\Delta x}{2}\right)}$$

Thus, ρ_{n+1}/ρ_n will remain less than 1 for all values of $\Delta t/(\Delta x)^2$.

9. PARABOLIC EQUATION WITH SMALL DISPERSION COEFFICIENT

Both the Crank-Nicolson and the backward difference equation become unsatisfactory under certain conditions if the coefficient of the second space derivative is small compared to the coefficient of the first space derivative. The differential equation to be considered is

$$\frac{\partial^2 u}{\partial x^2} - b\frac{\partial u}{\partial x} = \frac{\partial u}{\partial t} \tag{2-25}$$

This equation describes the one-dimensional flow of a fluid with dispersion or diffusion. The parameter b is the ratio of the velocity of flow to the dispersion coefficient. The first space derivative, in either of the finite difference equations being considered, is replaced by the difference over two increments centered at x_i. When the coefficient b is large, or the dispersion coefficient is small, this term is the controlling term on the left side of the equation. As a result, the first space derivative over two increments is set equal to the first time derivative over one increment. The value of $u_{i,n}$ has little influence from the spatial derivatives in determining the value of $u_{i,n+1}$. Consequently, the values of u at a given time level oscillate around the true curve. The extent of this oscillation depends on the relative values of b and Δx. Price *et al.* (*2*) have shown that, for the Crank-Nicolson equation, there will be no oscillation when

$$\frac{b\,\Delta x}{2} < 1 \tag{2-26}$$

for $0 \leq x \leq 1$.

For large values of b, small values of Δx must be used to eliminate this oscillation. Two methods have been developed for eliminating this oscillation. However, these are important mainly for problems in two space dimensions. For problems in only one space dimension, it should usually be possible to use a Δx small enough to eliminate the oscillation. Of course, when the dispersion coefficient becomes zero so that b becomes unbounded, and the equation is no longer parabolic, it is impossible to use a small-enough increment. Consequently, the Crank-Nicolson and backward difference equations are not applicable to the solution of differential equations of this type. Methods for the numerical solution of hyperbolic differential equations are the subject of the next chapter.

EXAMPLE 2-1. HEAT FLOW IN INSULATED ROD

$$\frac{\partial^2 u}{\partial x^2} = \frac{\partial u}{\partial t}$$

$$u(x, 0) = 0 \qquad \text{all } x$$

$$u(0, t) = 0 \qquad \text{all } t$$

$$u(1, t) = 1 \qquad \text{all } t$$

Define $x_i = i(\Delta x)$. Use Crank-Nicolson method.

Use 20 increments; so $\Delta x = 0.05$, $u_0 = 0$, $u_{20} = 1$.

The finite difference equations are:

For $2 \leq i \leq 18$:

$$u_{i-1,n+1} + \left[-2 - 2\frac{(\Delta x)^2}{\Delta t}\right]u_{i,n+1} + u_{i+1,n+1} = -u_{i-1,n} - u_{i+1,n} + \left[2 - 2\frac{(\Delta x)^2}{\Delta t}\right]u_{i,n}$$

For $i = 1$:

$$\left[-2 - 2\frac{(\Delta x)^2}{\Delta t}\right]u_{1,n+1} + u_{2,n+1} = \left[2 - 2\frac{(\Delta x)^2}{\Delta t}\right]u_{1,n} - u_{2,n}$$

For $i = 19$:

$$u_{18,n+1} + \left[-2 - 2\frac{(\Delta x)^2}{\Delta t}\right]u_{19,n+1} = -u_{18,n} + \left[2 - 2\frac{(\Delta x)^2}{\Delta t}\right]u_{19,n} - 2$$

In the tridiagonal system,

$$a_i = c_i = 1 \qquad \text{for all } i \text{ except } a_1 = c_{19} = 0$$

$$b_i = -2 - 2\frac{(\Delta x)^2}{\Delta t} \qquad \text{for all } i$$

Define,

$$b_i = BB = -2 - 2\frac{(\Delta x)^2}{\Delta t}$$

Then

$$\beta_i = BB - \frac{1}{\beta_{i-1}}$$

$$\gamma_i = \frac{d_i - \gamma_{i-1}}{\beta_i}$$

For the program, call $\beta_i = B(I)$, $\gamma_i = G(I)$.

Begin with $\Delta t = (\Delta x)^2 = 0.0025$.

Increase Δt by 10% each time step.

Program for 40 time steps.

The following program was compiled by WATFOR and executed with on-line output in 32 seconds on IBM 7044. Steady state was reached at 39th time step.

```
      DIMENSION U(20),B(20),G(20)
      DT=0.0025
      U0=0.
      U(20)=1.
      DO 1 I=1,19
    1 U(I)=0.
      T=0.
      N=0
      M=1
      WRITE (6,20)
      WRITE (6,21) N,T,DT
      WRITE (6,22) U0,(U(I),I=1,20)
      WRITE (6,23)
   20 FORMAT (1H1)
   21 FORMAT (1H0,I5,2F11.4)
   22 FORMAT (1H0,(11F11.4))
   23 FORMAT (1H )
      DO 2 N=1,40
      T=T+DT
      M=M+1
      BP= 0.005/DT
      BB= -2. - BP
      BD= 2.-BP
      B(1)=BB
      G(1)=(BD*U(1)-U(2))/BB
      DO 3 I=2,19
      B(I)=BB-1./B(I-1)
      D=-U(I-1) +BD*U(I)-U(I+1)
    3 G(I)=(D-G(I-1))/B(I)
      U(19)=G(19)-1./B(19)
      DO 4 J=1,18
      I=19-J
    4 U(I)=G(I)-U(I+1)/B(I)
      WRITE (6,21) N,T,DT
      WRITE (6,22) U0,(U(I),I=1,20)
      WRITE (6,23)
      IF (M-10) 2,5,5
    5 WRITE (6,20)
      M=0
    2 DT=1.1*DT
      CALL EXIT
      END
```

EXAMPLE 2-2. HEAT FLOW IN TAPERED ROD

$$\frac{\partial^2 u}{\partial x^2} + \frac{2}{x}\left(\frac{\partial u}{\partial x} - pu\right) = \frac{\partial u}{\partial t}$$

$$u(1, t) = 1 \qquad \text{all } t$$

$$\frac{\partial u}{\partial x} - pu = 0 \qquad \text{at } x = 0, \text{ all } t$$

$$u(x, 0) = 0 \qquad \text{all } x$$

Define $x_i = (i - \frac{1}{2})\,\Delta x$, so points are $\frac{1}{2}\,\Delta x$ from boundaries. Use Crank-Nicolson method.

Program for R increments, so $\Delta x = 1/R$.

The finite difference equations are:

For $2 \leq i \leq (R - 1)$:

$$\left(\frac{2i-3}{2i-1}\right)u_{i-1,n+1} + \left[-2 - \frac{4p(\Delta x)}{2i-1} - \frac{2(\Delta x)^2}{\Delta t}\right]u_{i,n+1} + \left(\frac{2i+1}{2i-1}\right)u_{i+1,n+1}$$

$$= -\left(\frac{2i-3}{2i-1}\right)u_{i-1,n} - \left(\frac{2i+1}{2i-1}\right)u_{i+1,n} + \left(2 + \frac{4p(\Delta x)}{2i-1} - \frac{2(\Delta x)^2}{\Delta t}\right)u_{i,n}$$

For $i = 1$:

$$\left[-2 - 4p(\Delta x) - \frac{2 - p(\Delta x)}{2 + p(\Delta x)} - \frac{2(\Delta x)^2}{\Delta t}\right]u_{1,n+1} + 3u_{2,n+1}$$

$$= -3u_{2,n} + \left[2 + 4p(\Delta x) + \frac{2 - p(\Delta x)}{2 + p(\Delta x)} - \frac{2(\Delta x)^2}{\Delta t}\right]u_{1,n}$$

For $i = R$:

$$\left(\frac{2R-3}{2R-1}\right)u_{R-1,n+1} + \left[-2 - \frac{4p(\Delta x)}{2R-1} - \frac{2R+1}{2R-1} - \frac{2(\Delta x)^2}{\Delta t}\right]u_{R,n+1}$$

$$= -\left(\frac{2R-3}{2R-1}\right)u_{R-1,n} + \left[2 + \frac{4p(\Delta x)}{2R-1} + \frac{2R+1}{2R-1} - \frac{2(\Delta x)^2}{\Delta t}\right]u_{R,n}$$

$$- 4\left(\frac{2R+1}{2R-1}\right)$$

In the tridiagonal system:

$$a_i = \frac{2i-3}{2i-1} \qquad \text{and} \qquad c_i = \frac{2i+1}{2i-1} \qquad \text{for } 1 \leq i \leq R$$

Actually $a_1 = c_R = 0$, but these are not used in the algorithm, and the defined values will be used elsewhere.

The part of b_1 independent of Δt will be defined as $b_i = 2 + [4p(\Delta x)/(2i - 1)]$ with variations for b_i and b_R. The algorithm can then be programmed essentially as it is given, with $2(\Delta x)^2/\Delta t$ added to each b_i.

Call $\beta_i = BT(I)$ and $\gamma_i = G(I)$

Since the limit of the i index is fixed point, call $R = IR$.

Program for NT, total time steps: The number of time steps printed per page (MM) from the table is:

IR less than:	11	21	31	41	61	81	101
MM is:	12	10	8	7	6	5	4

This program was run for $IR = 20$, $p = 10$. Steady state was reached after 25 time steps. With WATFOR compiler on an IBM 7044, the running time was 28 seconds for 40 time steps.

```
      DIMENSION U(100), A(100), B(100), C(100), BT(100), G(100)
      READ (5,30) P, IR, MM, NT
   30 FORMAT (E10.4, 3I3)
      IRM1 = IR - 1
      R = IR
      DX = 1./R
      PDX = P * DX
      DO 1 I=1,IR
      U(I) = 0.0
      FI = 2 * I
      A(I) = (FI-3.) / (FI-1.)
      B(I) = 2. + (4.*PDX) / (FI-1.)
    1 C(I) = (FI+1.) / (FI-1.)
      B(1) = B(1) - A(1) * (2.-PDX) / (2.+PDX)
      B(IR) = B(IR) + C(IR)
      DT = DX * DX
      T = 0.0
      N = 0
      M = 1
      WRITE (6,20)
      WRITE (6,24) P
      WRITE (6,21) N, T, DT
      WRITE (6,22) (U(I), I=1,IR)
      WRITE (6,23)
   20 FORMAT (1H1)
   21 FORMAT (1H0, I5, 2F11.4)
   22 FORMAT (1H0, (10F11.4))
   23 FORMAT (1H )
   24 FORMAT (1H ,F11.4)
      DO 2 N=1,NT
      T = T + DT
      M = M + 1
      DXT = (2.*DX*DX) / DT
      BT(1) = -B(1) - DXT
      G(1) =(-C(1) * U(2) + (B(1) - DXT) * U(1)) / BT(1)
      DO 3 I=2,IRM1
      BT(I) = -B(I) - DXT - A(I) * C(I-1) / BT(I-1)
      D = -A(I) * U(I-1) - C(I) * U(I+1) + (B(I) - DXT) * U(I)
    3 G(I) = (D-A(I) * G(I-1)) / BT(I)
      BT(IR) = -B(IR) - DXT - A(IR) * C(IRM1) / BT(IRM1)
      D = -A(IR) * U(IRM1) + (B(IR) - DXT) * U(IR) - 4.*C(IR)
      U(IR) = (D - A(IR) * G(IRM1) ) / BT(IR)
      DO 4 J=1,IRM1
      I = IR -J
    4 U(I) = G(I) - C(I) * U(I+1) / BT(I)
      WRITE (6,21) N, T, DT
      WRITE (6,22) (U(I), I=1,IR)
      WRITE (6,23)
      IF (M-MM) 2, 5, 5
    5 WRITE (6,20)
      M = 0
    2 DT = 1.1 * DT
      CALL EXIT
      END
```

Chapter Three

LINEAR HYPERBOLIC PARTIAL DIFFERENTIAL EQUATIONS

1. INTRODUCTION

One-dimensional hyperbolic differential equations arise from pure convection problems. The simplest equation describes the flow of a fluid through a tube with no transfer of the quantity conserved and with no generation or consumption. This equation is

$$-b\frac{\partial u}{\partial x} = \frac{\partial u}{\partial t} \tag{3-1}$$

The centered difference equation is used for its numerical solution.

2. THE CENTERED DIFFERENCE EQUATION

For the centered difference equation, the finite difference analogs are centered in both space and time with respect to the grid points at which the values of the dependent variable are determined. In Figure 3-1 two successive time levels of grid points are shown. The grid points at which the dependent variable, u, is to be computed are represented by the circles. The space index, i, and the time index, n, take on integral values at these points. A single, representative point about which the finite difference analogs are written is shown; this point is designated by a cross. The coordinate of this point is $x_{i-1/2}$, $t_{n+1/2}$. As in the Crank-Nicolson equation, the space derivative at the $n + 1/2$ time level is approximated as the average of the space derivatives at the time levels for t_n and t_{n+1}. This analog is

$$\left(\frac{\partial u}{\partial x}\right)_{i-1/2,n+1/2} \approx \frac{1}{2}\left(\frac{u_{i,n+1} - u_{i-1,n+1}}{\Delta x} + \frac{u_{i,n} - u_{i-1,n}}{\Delta x}\right) \tag{3-2}$$

It can be shown that this analog is second-order correct.

In a similar manner, the time derivative at the $i - 1/2$ space position is approximated as the average of the time derivatives at the space positions for

x_i and x_{i-1}. This derivative is also second-order correct, and it is

$$\left(\frac{\partial u}{\partial t}\right)_{i-1/2,n+1/2} \approx \frac{1}{2}\left(\frac{u_{i,n+1} - u_{i,n}}{\Delta t} + \frac{u_{i-1,n+1} - u_{i-1,n}}{\Delta t}\right) \tag{3-3}$$

Both of these analogs contain values of the dependent variable at the same four points in the grid. Consequently, when they are substituted into equation (3-1), the resulting finite difference equation will contain values of u at these four points. This equation is

$$\left(\frac{b}{\Delta x} + \frac{1}{\Delta t}\right)u_{i,n+1} = \left(\frac{b}{\Delta x} - \frac{1}{\Delta t}\right)(u_{i-1,n+1} - u_{i,n}) + \left(\frac{b}{\Delta x} + \frac{1}{\Delta t}\right)u_{i-1,n} \tag{3-4}$$

Wendroff (*3*) has shown that this equation is stable for any ratio of $\Delta x/\Delta t$.

t_{n+1}, t_n, x_0, x_{i-1}, x_i

Figure 3-1. Center of analogs for centered difference equation.

The centered difference analog of equation (3-1) is explicit. For a typical set of initial and boundary conditions, u will be known for all values of x at $t = 0$ and for all values of t at $x = 0$. When equation (3-4) is written for $i = 1$, $n = 0$, then $u_{0,0}$, $u_{0,1}$, and $u_{1,0}$ will be known so the value of $u_{1,1}$ can be obtained directly. The same equation can be used successively to compute $u_{2,1}$, $u_{3,1}$, etc. When enough values at the time level for $n = 1$ are obtained, the equation can be used to obtain the values at $n = 2$, and this procedure can be continued for as many time levels as desired. It is interesting to note that the values of u at the first space position could be computed for as many time levels as desired before any values at the second space position were computed.

3. TRUNCATION ERROR OF CENTERED DIFFERENCE EQUATION

Although the centered difference equation is stable for any ratio of increments, the truncation error can be minimized by a proper choice of this ratio. In fact, there is no truncation error at all in equation (3-4) for the proper value of the ratio $\Delta x/\Delta t$. This value is b, the velocity of the fluid.

An examination of the complete expressions for the derivatives will demonstrate this fact. The complete expressions for the derivatives given in equations (3-2) and (3-3) are

$$\left(\frac{\partial u}{\partial x}\right)_{i-\frac{1}{2},n+\frac{1}{2}} = \frac{1}{2}\left(\frac{u_{i,n+1} - u_{i-1,n+1}}{\Delta x} + \frac{u_{i,n} - u_{i-1,n}}{\Delta x}\right) - \sum_{n=1}^{\infty}\sum_{r=0}^{n}\left(\frac{\Delta x}{2}\right)^{2r}\left(\frac{\Delta t}{2}\right)^{(2n-2r)}\frac{1}{(2r+1)!\,(2n-2r)!} \times \left[\frac{\partial^{(2n+1)}u}{\partial x^{(2r+1)}\,\partial t^{(2n-2r)}}\right]_{i-\frac{1}{2},n+\frac{1}{2}} \tag{3-2a}$$

and

$$\left(\frac{\partial u}{\partial t}\right)_{i-\frac{1}{2},n+\frac{1}{2}} = \frac{1}{2}\left(\frac{u_{i,n+1} - u_{i,n}}{\Delta t} + \frac{u_{i-1,n+1} - u_{i-1,n}}{\Delta t}\right) - \sum_{n=1}^{\infty}\sum_{r=0}^{n}\left(\frac{\Delta x}{2}\right)^{(2n-2r)}\left(\frac{\Delta t}{2}\right)^{2r}\frac{1}{(2r+1)!\,(2n-2r)!} \times \left[\frac{\partial^{(2n+1)}u}{\partial x^{(2n-2r)}\,\partial t^{(2r+1)}}\right]_{i-\frac{1}{2},n+\frac{1}{2}} \tag{3-3a}$$

When equation (3-4) is subtracted from equation (3-1), the resulting equation of the error terms is

$$-b\sum_{n=1}^{\infty}\sum_{r=0}^{n}\left(\frac{\Delta x}{2}\right)^{2r}\left(\frac{\Delta t}{2}\right)^{(2n-2r)}\frac{1}{(2r+1)!\,(2n-2r)!}\left[\frac{\partial^{(2n+1)}u}{\partial x^{(2r+1)}\,\partial t^{(2n-2r)}}\right]_{i-\frac{1}{2},n+\frac{1}{2}} = \sum_{n=1}^{\infty}\sum_{r=0}^{n}\left(\frac{\Delta x}{2}\right)^{(2n-2r)}\left(\frac{\Delta t}{2}\right)^{2r}\frac{1}{(2r+1)!\,(2n-2r)!}\left[\frac{\partial^{(2n+1)}u}{\partial x^{(2n-2r)}\,\partial t^{(2r+1)}}\right]_{i-\frac{1}{2},n+\frac{1}{2}} \tag{3-5}$$

When equation (3-1) is differentiated successively with respect to x and t, it can be shown that the higher derivatives are related by

$$(-b)^{(r-s)}\frac{\partial^n u}{\partial x^r\,\partial t^{(n-r)}} = \frac{\partial^n u}{\partial x^s\,\partial t^{(n-s)}} \qquad \text{for all } n, \text{ for } 0 \le r \le n, \; 0 \le s \le n \tag{3-6}$$

From this relation it can be readily shown that the corresponding terms of equation (3-5) cancel when $(\Delta x/\Delta t) = b$, and there is no truncation error when equation (3-4) is used to represent equation (3-1). The results obtained from using this increment ratio in equation (3-4) are found to correspond exactly to the physical model.

With $\Delta x/\Delta t = b$, equation (3-4) becomes

$$u_{i,n+1} = u_{i-1,n} \tag{3-4a}$$

and the value of the dependent variable is moved forward one space increment for each new time increment. For the physical model there is no mechanism for dispersion of the quantity conserved, and there is no process by which it is added or removed. Thus, the quantity measured by the dependent variable, u, simply moves down the tube with velocity b; and in the time increment Δt, the value of u at a given position has moved downstream a distance $b\,\Delta t$ or Δx. The numerical solution therefore, agrees exactly with the physical model.

4. TYPES OF EQUATIONS SOLVED BY CENTERED DIFFERENCE METHOD

Of course, numerical methods are not required to solve hyperbolic equations as simple as equation (3-1). For many physical problems, however, the material conserved is being added to or removed from the stream by a mechanism such as adsorption, heat transfer, or chemical reaction. For these equations a numerical solution is often desired, and the centered difference equation is very satisfactory. The finite difference equations are explicit; and, since the differences are all centered, they should be stable for almost any ratio of increments. Also, a reasonably large increment size can be used because the finite differences are second-order correct. These equations will not have zero truncation error for a ratio of $\Delta x/\Delta t$ equal to the fluid velocity, but this ratio should give a small truncation error in most cases.

A large number of physical problems are described by two or more coupled hyperbolic equations. Some of these arise from fluid-to-fluid heat exchangers, while others describe fixed-bed adsorbers. In these latter cases, there is no spatial derivative in the equation for the solid, since it is not moving. The centered difference equations have proved excellent for these problems also. When all the boundary conditions are known at the same point in space, the two differential equations yield two finite difference equations for each point in space. Only one unknown value of each dependent variable appears in these two equations; so the method can be readily made explicit. For the very important mixed boundary value problems, where the boundary conditions are given at different points in space, the centered difference method has proved to be extremely valuable. There are a number of complications which arise; therefore this case merits further discussion.

5. SPLIT BOUNDARY VALUE PROBLEMS

It is most convenient to discuss this case by considering a simple physical problem. Let us consider the case of an unsteady-state, countercurrent heat exchanger. The equations describing this problem are

$$-b_1\left(\frac{\partial u}{\partial x}\right) - c_1(u - v) = \left(\frac{\partial u}{\partial t}\right) \tag{3-7}$$

$$b_2\left(\frac{\partial v}{\partial x}\right) + c_2(u - v) = \left(\frac{\partial v}{\partial t}\right) \tag{3-8}$$

Equation (3-7) is a heat balance on the fluid flowing in the direction of increasing length; the dimensionless temperature of this fluid is the dependent variable, u. Equation (3-8) is a heat balance on the fluid flowing countercurrently in the direction of decreasing length, and its dimensionless temperature is designated as v. The parameters b_1 and b_2 are dimensionless velocities of the two streams, and the parameters c_1 and c_2 contain the coefficient of heat transfer between the two flowing streams. The terms containing $(u - v)$ are thus the terms which account for the transfer of heat between the two flowing streams. The heat capacity of the tube wall between the two streams is not included in this simple model. It is assumed in this problem that the size of the exchanger and the various parameters are known.

Initial conditions and boundary conditions must be known for this problem. The initial conditions are simply a specification of the temperatures at all positions in the exchanger at the initial time. For a real problem, the inlet temperatures of the two streams are specified. The input temperature of one fluid is known at one end of the exchanger, and that of the other fluid is known at the opposite end. A simple set of boundary conditions is

$$\begin{aligned} u(x, 0) &= 0 \qquad \text{for } x > 0 \\ u(0, 0) &= 1 \\ v(x, 0) &= 0 \\ u(0, t) &= 1 \\ v(1, t) &= 0 \end{aligned} \tag{3-9}$$

The resulting problem is similar to that described by equations (2-1), (2-2), and (2-3) in that the boundary conditions at both ends are introduced at a single time level. The centered difference equations are, consequently, implicit. These equations are not of such a form that they can be solved by the Thomas algorithm. However, Douglas *et al.* (*4*) have developed an algorithm for a system of equations which are tridiagonal in two dependent variables, and this algorithm can be used to solve the centered finite difference equations for the split boundary value problem. In order that the nomenclature and subscripts of the difference equations take a form that is amenable to ready solution by this algorithm, the position index has a different value for u than it has for v at the same point. The nomenclature and indexing are illustrated in Figure 3-2. The circles designate points at which values of the dependent

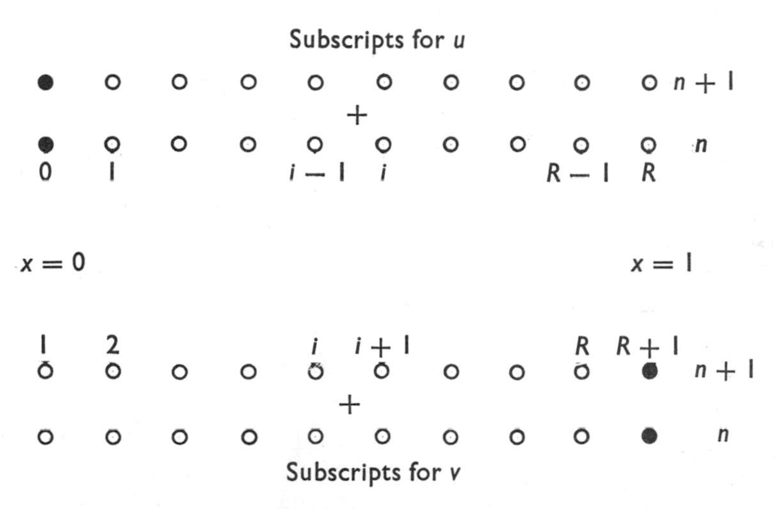

Figure 3-2. Grid points for split boundary conditions.

variables are computed. The four shaded circles designate the points at which values of the dependent variables are given by the boundary conditions. The dependent variable u is known at the position where $x = 0$; thus the first unknown value of u is at the position where $x = \Delta x$. The index i takes the value of 1 at this point for the variable u; it takes the value R for the last unknown value of u at the position where $x = 1$. The dependent variable v is known at the position where $x = 1$. Consequently, the first unknown value of v is at the position where $x = 0$, and the index i takes the value of 1 at this point for the variable v. It takes the value R at the last unknown value of v at the position where $x = 1 - \Delta x$. As a result, the position index at a given point has a value one greater for v than it has for u.

A typical set of equations are formulated about the point designated by the cross in Figure 3-2. The centered differences corresponding to those given in equations (3-2) and (3-3) are used for the derivatives. The average of the values of the dependent variable at the four adjacent points denoted by circles is used to represent the value of the dependent variable at the point denoted by the cross. The resulting finite difference equations are

$$\left(-\frac{b_1}{\Delta x}+\frac{c_1}{2}+\frac{1}{\Delta t}\right)u_{i-1,n+1}+\left(\frac{b_1}{\Delta x}+\frac{c_1}{2}+\frac{1}{\Delta t}\right)u_{i,n+1}+\left(-\frac{c_1}{2}\right)v_{i,n+1}$$

$$+\left(-\frac{c_1}{2}\right)v_{i+1,n+1}=\left(\frac{b_1}{\Delta x}-\frac{c_1}{2}+\frac{1}{\Delta t}\right)u_{i-1,n}+\left(-\frac{b_1}{\Delta x}-\frac{c_1}{2}+\frac{1}{\Delta t}\right)u_{i,n}$$

$$+\frac{c_1}{2}(v_{i+1,n}+v_{i,n}) \quad (3\text{-}10)$$

and

$$\left(\frac{c_2}{2}\right)u_{i-1,n+1}+\left(\frac{c_2}{2}\right)u_{i,n+1}+\left(-\frac{b_2}{\Delta x}-\frac{c_2}{2}-\frac{1}{\Delta t}\right)v_{i,n+1}+\left(\frac{b_2}{\Delta x}-\frac{c_2}{2}-\frac{1}{\Delta t}\right)\cdot$$

$$\cdot v_{i+1,n+1}=-\frac{c_2}{2}(u_{i,n}+u_{i-1,n})+\left(\frac{b_2}{\Delta x}+\frac{c_2}{2}-\frac{1}{\Delta t}\right)v_{i,n}$$

$$+\left(-\frac{b_2}{\Delta x}+\frac{c_2}{2}-\frac{1}{\Delta t}\right)v_{i+1,n} \quad (3\text{-}11)$$

The boundary equations for $i = 1$ and $i = R$ will be slightly different, since $u_{0,n+1}$ and $v_{R+1,n+1}$ are known from the boundary conditions. Notice that, although only two unknown values of each of the dependent variables appear in an equation, there are three values of the position index in the equation. Consequently, the equations are a simplified form of

$$a_i^{(1)}u_{i-1}+a_i^{(2)}v_{i-1}+b_i^{(1)}u_i+b_i^{(2)}v_i+c_i^{(1)}u_{i+1}+c_i^{(2)}v_{i+1}=d_i^{(1)} \quad (3\text{-}10a)$$

and

$$a_i^{(3)}u_{i-1}+a_i^{(4)}v_{i-1}+b_i^{(3)}u_i+b_i^{(4)}v_i+c_i^{(3)}u_{i+1}+c_i^{(4)}v_{i+1}=d_i^{(2)} \quad (3\text{-}11a)$$

with

$$a_1^{(1)}=a_1^{(2)}=a_1^{(3)}=a_1^{(4)}=c_R^{(1)}=c_R^{(2)}=c_R^{(3)}=c_R^{(4)}=0$$

For the special case of equations (3-10) and (3-11), four additional coefficients in each set of equations are zero. These are $a_i^{(2)}$, $a_i^{(4)}$, $c_i^{(1)}$, and $c_i^{(3)}$. The complete algorithm for solving equations (3-10a) and (3-11a) is given in the Appendix. This algorithm can be simplified considerably for equations (3-10) and (3-11).

The centered difference method has been used quite successfully by Herron and von Rosenberg (*5*) for the solution of equations describing unsteady-state, countercurrent heat exchangers. The method is rapid, and the solution can be obtained on a small or medium-sized computer. The numerical solution agreed very closely with the one analytical solution which was available even for a relatively large space increment. Since there are two velocities in this problem—namely, b_1 and b_2 in equations (3-7) and (3-8)—a choice must be for the ratio $\Delta x/\Delta t$. For the boundary conditions of equation (3-9), a step function is imposed on the dependent variable u at $x = 0$. As a result, the profile of u will be much less smooth than that for v; thus it is more important to minimize the truncation error in u, and the ratio used was $\Delta x/\Delta t = b_1$. The results of the numerical solution demonstrated that this was the best ratio to use.

For a step change or finite discontinuity in the temperature of one stream, the results of equations (3-10) and (3-11) contain an oscillation in the neighborhood of the discontinuity. This oscillation causes no further error in the solution after the discontinuity has passed out of the exchanger, but it is desirable to eliminate the oscillation. A recent modification of the analog to the perturbed temperature—u in the example problem—completely eliminates the oscillation. In the interphase transfer terms $(u - v)$ the value of u at the center point is represented by the average of u at the two points on the diagonal in the direction of flow; thus, $u_{i-\frac{1}{2},n+\frac{1}{2}} \approx \frac{1}{2}(u_{i,n+1} + u_{i-1,n})$. The resulting finite difference equations are actually a little simpler than are equations (3-10) and (3-11), but they are also of the form of equations (3-10a) and (3-11a). This diagonal analog is recommended when there is a step change in one of the dependent variables.

The set of finite difference equations given by equations (3-10) and (3-11) for all values of the index i can be rearranged so that the coefficient matrix is of the pentadiagonal form. In this matrix the only nonzero terms are the diagonal and two terms in each row to the right and to the left of the diagonal. An algorithm for the solution of equations of this form has been developed and is reported by Blair (*6*). This algorithm is presented in the Appendix. An alternate method is thus obtained for solving the finite difference equations for split boundary value problems of hyperbolic equations in two dependent variables. For problems involving more than two dependent variables it is quite important that various arrangements of the equations be obtained so that the most efficient algorithm can be used for their solution. Consequently, various arrangements of the equations for a number of problems are presented in Chapter Four. The rearrangement of equations (3-10) and (3-11) to fit the pentadiagonal form is included.

6. SETS OF PARABOLIC AND HYPERBOLIC EQUATIONS

The model of the countercurrent heat exchanger given by equations (3-7) and (3-8) does not include the effect of the heat capacity of the tube which separates the streams. The set of equations which results when this effect is included consists of two hyperbolic equations and one parabolic equation. These equations are

$$-b_1 \frac{\partial u}{\partial x} - c_1(u - w) = \frac{\partial u}{\partial t} \tag{3-12}$$

$$a_1 \frac{\partial^2 w}{\partial x^2} + c_2(u - w) - c_3(w - v) = \frac{\partial w}{\partial t} \tag{3-13}$$

$$b_2 \frac{\partial v}{\partial x} + c_4(w - v) = \frac{\partial v}{\partial t} \tag{3-14}$$

This set of equations now also includes the temperature of the tube, w, and the heat balance on the tube is included as equation (3-13). Conduction is included in this equation, with the parameter a_1 representing the dimensionless conductivity. These equations can be solved by using the Crank-Nicolson method for equation (3-13) and the centered difference method for equations (3-12) and (3-14). In order that the equations can be set up in the simplest form, the points for which u and v are determined are placed on each boundary. The indexing for these variables is that in Figure 3-2. The points at which w is determined are shifted one-half increment from the boundary, and the x index for these points is used so that

$$x_i = (i - \tfrac{1}{2})\,\Delta x \tag{1-19}$$

This arrangement of points is shown in Figure 3-3. The finite difference analogs for the u and v terms given above can be used in the finite difference equations. The Crank-Nicolson analogs of equations (2-14) and (2-15) can be used for the derivatives of w. An analog for $w_{i,n+1/2}$ is thus needed, and it is

$$w_{i,n+1/2} \approx \tfrac{1}{2}(w_{i,n} + w_{i,n+1}) \tag{3-15}$$

The resulting equations are a special case of a matrix which is tridiagonal in three dependent variables. Herron (7) has developed an algorithm for the

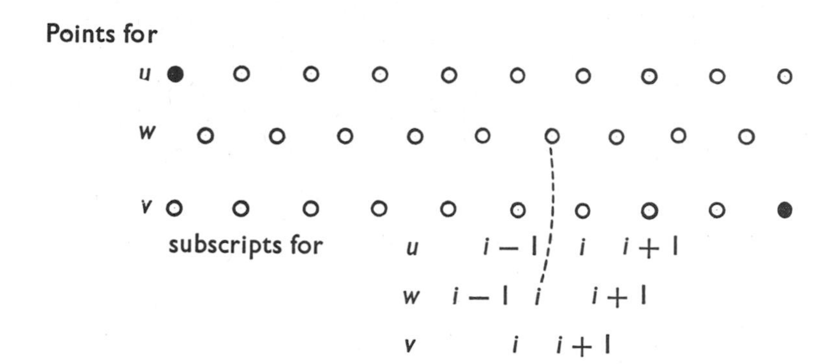

Figure 3-3. Grid points for parabolic-hyperbolic problem.

solution of this system of equations; this algorithm is presented in the Appendix. The method of handling these three equations was first presented by Herron and von Rosenberg (*5*).

7. SETS OF HYPERBOLIC EQUATIONS WITH MORE THAN TWO DEPENDENT VARIABLES

In a number of hyperbolic systems of interest there are more than two dependent variables. For the two-tube pass, single-shell pass heat exchanger there are three dependent variables, and more complicated arrangements of heat exchangers must be described by even more. Three dependent variables must be used to describe transient, compressible flow problems. These processes all result in split boundary conditions. Thus, when the centered difference method is used, another special case of the tri-tridiagonal system results, and the algorithm mentioned above can be used for their solution. For systems of more than three-dependent variables, there is a possibility that some of the finite difference equations can be uncoupled so that only three sets of dependent variables need to be determined simultaneously, and the tri-tridiagonal algorithm will be sufficient. No algorithm for a set of four dependent variables with tri-diagonal matrices has been derived. However, this set of equations can be solved by a general algorithm for a matrix of the band form with suitable rearrangement of the coefficients. The rearrangement for any size tridiagonal system and the general band algorithm are discussed in the next chapter.

EXAMPLE 3-1. COUNTERCURRENT HEAT EXCHANGER

$$-\frac{\partial u}{\partial x} - c_1(u - v) = \frac{\partial u}{\partial t}$$

$$b_2 \frac{\partial v}{\partial x} + c_2(u - v) = \frac{\partial v}{\partial t}$$

$$\begin{array}{ll} u(x, 0) = 0 & x > 0 \\ u(0, 0) = 1 & \\ v(x, 0) = 0 & \text{all } x \\ \left.\begin{array}{l} u(0, t) = 1 \\ v(1, t) = 0 \end{array}\right\} & \text{all } t \end{array}$$

Define $x_i = i(\Delta x)$ for u, and $x_i = (i - 1)\,\Delta x$ for v.

Use centered difference method. The coefficient of $\partial u/\partial x$ is 1, and u is the perturbed variable, so make $\Delta x = \Delta t$ to minimize truncation.

Program for R increments; so $\Delta x = 1/R$.

The finite difference equations are:

For $2 \leq i \leq R - 1$:

$$u_{i-1,n+1} + \left[\frac{4}{c_1(\Delta x)} + 1\right] u_{i,n+1} - v_{i,n+1} - v_{i+1,n+1}$$

$$= \left[\frac{4}{c_1(\Delta x)} - 1\right] u_{i-1,n} - u_{i,n} + v_{i,n} + v_{i+1,n}$$

and

$$u_{i-1,n+1} + u_{i,n+1} + \left[-\frac{2(b_2 + 1)}{c_2(\Delta x)} - 1\right] v_{i,n+1} + \left[\frac{2(b_2 - 1)}{c_2(\Delta x)} - 1\right] v_{i+1,n+1}$$

$$= -u_{i-1,n} - u_{i,n} + \left[\frac{2(b_2 - 1)}{c_2(\Delta x)} + 1\right] v_{i,n} + \left[-\frac{2(b_2 + 1)}{c_2(\Delta x)} + 1\right] v_{i+1,n}$$

The equations for $i = 1$ are the same except that $u_{0,n+1} = 1$ is moved to the right side of the equations, and $u_{0,n} = 1$ also.

The equations for $i = R$ are the same except that $v_{R+1,n+1} = 0$ is moved to right side of the equations, and $v_{R+1,n} = 0$ also.

In the bi-tridiagonal system:

$$a_i^{(2)} = a_i^{(4)} = c_i^{(1)} = c_i^{(3)} = 0 \qquad \text{for all } i$$

$$a_i^{(1)} = a_i^{(3)} = b_i^{(3)} = 1 \qquad \text{and} \qquad c_i^{(2)} = b_i^{(2)} = -1 \qquad \text{for all } i$$

Also

$$b_i^{(1)} = 1 + \frac{4}{c_1(\Delta x)} = B1$$

$$b_i^{(4)} = -1 - \frac{2(b_2 + 1)}{c_2(\Delta x)} = B4$$

$$c_i^{(4)} = \left[\frac{2(b_2 - 1)}{c_2(\Delta x)} - 1\right] = C4$$

Also, define

$$B5 = \frac{4}{c_1(\Delta x)} - 1$$

$$B6 = \frac{2(b_2 - 1)}{c_2(\Delta x)} + 1$$

$$C6 = 1 - \frac{2(b_2 + 1)}{c_2(\Delta x)}$$

For the program, call the three coefficients in the differential equations $CC1$, $CC2$, and $BB2$.

Call the number of increments IR.

Use a simplified algorithm, calling $\beta_i^{(2)} = BT2$, $\beta_i^{(4)} = BT4$, $\lambda_i^{(2)} = EL2(I)$, $\lambda_i^{(4)} = EL4(I)$, $\mu_i = EM$, $\gamma_i^{(1)} = G1(I)$ and $\gamma_i^{(2)} = G2(I)$.

The simplified algorithm is

$$\beta_i^{(2)} = -1 - \lambda_{i-1}^{(2)};\ \beta_i^{(4)} = B4 - \lambda_{i-1}^{(2)}$$

$$\mu_i = b_i^{(1)} \cdot \beta_i^{(4)} - \beta_i^{(2)}$$

$$\lambda_i^{(2)} = (-\beta_i^{(4)} - c_i^{(4)} \cdot \beta_i^{(2)})/\mu_i$$

$$\lambda_i^{(4)} = (b_i^{(1)} \cdot c_i^{(4)} + 1)/\mu_i$$

$$\gamma_i^{(1)} = [\beta_i^{(4)}(d_i^{(1)} - \gamma_{i-1}^{(1)}) - \beta_i^{(2)}(d_i^{(2)} - \gamma_{i-1}^{(1)})]/\mu_i$$

$$\gamma_i^{(2)} = [b_i^{(1)}(d_i^{(2)} - \gamma_{i-1}^{(1)}) - (d_i^{(1)} - \gamma_{i-1}^{(1)})]/\mu_i$$

The back solution is

$$u_R = \gamma_R^{(1)} \quad \text{and} \quad v_R = \gamma_R^{(2)}$$

Then

$$u_i = \gamma_i^{(1)} - \lambda_i^{(2)} v_{i+1} \quad \text{and} \quad v_i = \gamma_i^{(2)} - \lambda_i^{(4)} v_{i+1}$$

The number of time steps printed per page (MM) from the table is

IR less than:	11	21	31	51	71	101
MM is:	8	6	5	4	3	2

This program was run for $c_1 = 0.05$, $c_2 = 0.05$, $b_2 = 0.5$, $IR = 20$. With WATFOR compiler on an IBM 7044, the running time was 60 seconds for 60 time steps.

```
      DIMENSION U(101), EL2(101), EL4(101), G1(101), G2(101), V(101)
      READ (5,30) CC1,CC2,BB2,IR,MM,NT
30    FORMAT (3E10.4,3I3)
      R = IR
      DX = 1./R
      T1 = 4. * R / CC1
      B1 = 1. + T1
      B5 = T1 - 1.
      T2 = 2. * R * (BB2+1.) / CC2
      B4 = -1. - T2
      C6 = 1. - T2
      T3 = 2. * R * (BB2-1.) / CC2
      C4 = T3 - 1.
      B6 = T3 + 1.
      DT = DX
      T = 0.0
      N = 0
      U0 = 1.
      IRM1 = IR - 1
      IRP1 = IR + 1
      DO 1 I=1,IRP1
      U(I) = 0.0
1     V(I) = 0.0
      ELT = B1 * C4 + 1.
      M = 1
      WRITE (6,20)
      WRITE (6,21) N,T
      WRITE (6,22) U0, (U(I), I=1,IR)
      WRITE (6,22) (V(I), I=1,IRP1)
      WRITE (6,23)
20    FORMAT (1H1)
21    FORMAT (1H0, I5, F11.4)
22    FORMAT (1H0, (11F11.4))
23    FORMAT (1H )
      DO 2 N=1,NT
      T = T + DT
      M = M + 1
      EM = B1 * B4 - 1.
      EL2(1) = (C4-B4) / EM
      EL4(1) = ELT / EM
      D1 = -1. + B5 - U(1) + V(2) + V(1)
      D2 = -2. - U(1) + B6 * V(1) + C6 * V(2)
      G1(1) = (B4 * D1 + D2) / EM
      G2(1) = (B1 * D2 - D1) / EM
      DO 3 I=2,IR
      BT2 = -1. - EL2(I-1)
      BT4 = B4 - EL2(I-1)
      EM = B1 * BT4 - BT2
      EL2(I) = (-BT4 - C4 * BT2) / EM
      EL4(I) = ELT/EM
      D1 = B5 * U(I-1) - U(I) + V(I+1) + V(I)
      D2 = -U(I) - U(I-1) + B6 * V(I) + C6 * V(I+1)
      G1(I) = (BT4 * (D1 - G1(I-1)) - BT2 * (D2 - G1(I-1))) / EM
3     G2(I) = (B1 * (D2 - G1(I-1)) - D1 + G1(I-1)) / EM
      U(IR) = G1(IR)
      V(IR) = G2(IR)
      DO 4 J=1,IRM1
      I = IR - J
      U(I) = G1(I) - EL2(I) * V(I+1)
4     V(I) = G2(I) - EL4(I) * V(I+1)
      WRITE (6,21) N,T
      WRITE (6,22) U0, (U(I),I=1,IR)
      WRITE (6,22) (V(I), I=1,IRP1)
      WRITE (6,23)
      IF (M-MM) 2,5,5
5     WRITE (6,20)
      M = 0
2     CONTINUE
      CALL EXIT
      END
```

Chapter Four

ALTERNATE FORMS OF COEFFICIENT MATRICES

1. INTRODUCTION

It was observed in the previous chapter that the finite difference equations which are used to represent partial differential equations are often linear, algebraic equations. These equations can be arranged in a number of ways, and often one particular arrangement is most convenient, as it can be solved by an algorithm which is already available. The band form is particularly convenient, as an algorithm for its solution is available. In particular, it was observed that the equations arising from a two-variable, pure convective problem with split boundary conditions could be arranged into two common forms. An examination of these equations is an appropriate beginning.

2. EQUATIONS FOR SPLIT BOUNDARY VALUE, CONVECTIVE PROBLEM

The finite difference equations which arise in this case are simplified forms of equations (3-10a) and (3-11a). In matrix form they are

$$\begin{bmatrix} b_1^{(1)} & 0 & \cdots & & & & \\ a_2^{(1)} & b_2^{(1)} & 0 & \cdots & & & \\ & & \cdots & & & & \\ \cdots & 0 & a_i^{(1)} & b_i^{(1)} & 0 & \cdots & \\ & & & \cdots & & & \\ & & \cdots & 0 & a_R^{(1)} & b_R^{(1)} \end{bmatrix} \begin{bmatrix} u_1 \\ u_2 \\ \cdots \\ u_i \\ \cdots \\ u_R \end{bmatrix} + \begin{bmatrix} b_1^{(2)} & c_1^{(2)} & 0 & \cdots & & & \\ 0 & b_2^{(2)} & c_2^{(2)} & 0 & \cdots & & \\ & & \cdots & & & & \\ & \cdots & 0 & b_i^{(2)} & c_i^{(2)} & 0 & \\ & & & \cdots & & & \\ & & & \cdots & 0 & b_R^{(2)} \end{bmatrix} \begin{bmatrix} v_1 \\ v_2 \\ \cdots \\ v_i \\ \cdots \\ v_R \end{bmatrix} = \begin{bmatrix} d_1^{(1)} \\ d_2^{(1)} \\ \cdots \\ d_i^{(1)} \\ \cdots \\ d_R^{(1)} \end{bmatrix} \tag{4-1}$$

and

$$
\begin{bmatrix}
b_1^{(3)} & 0 & \cdots & & & & \\
a_2^{(3)} & b_2^{(3)} & 0 & \cdots & & & \\
 & & \cdots & & & & \\
\cdots & 0 & a_i^{(3)} & b_i^{(3)} & 0 & \cdots & \\
 & & & \cdots & & & \\
 & & & \cdots & 0 & a_R^{(3)} & b_R^{(3)}
\end{bmatrix}
\begin{bmatrix} u_1 \\ u_2 \\ \cdots \\ u_i \\ \cdots \\ u_R \end{bmatrix}
+
\begin{bmatrix}
b_1^{(4)} & c_1^{(4)} & 0 & \cdots & & & \\
0 & b_2^{(4)} & c_2^{(4)} & 0 & \cdots & & \\
 & & \cdots & & & & \\
 & \cdots & 0 & b_i^{(4)} & c_i^{(4)} & 0 & \\
 & & \cdots & & & & \\
 & & & \cdots & & 0 & b_R^{(4)}
\end{bmatrix}
\begin{bmatrix} v_1 \\ v_2 \\ \cdots \\ v_i \\ \cdots \\ v_R \end{bmatrix}
=
\begin{bmatrix} d_1^{(2)} \\ d_2^{(2)} \\ \cdots \\ d_i^{(2)} \\ \cdots \\ d_R^{(2)} \end{bmatrix}
\qquad (4\text{-}2)
$$

Of course, these two matrix equations, each with two variable vectors, can be written as a single matrix equation, and the single coefficient matrix can be rearranged by the rules for matrix manipulation. In this particular case, however, it is probably simpler to reorder the individual algebraic equations. The equations written as equations (4-1) and (4-2) are in a form such that they can be solved by the algorithm for the bi-tridiagonal matrix given in the Appendix.

These equations can also be arranged for solution by the algorithm for the pentadiagonal matrix given in the Appendix. The unknowns in the individual equations are rearranged into the order of u_{i-1}, v_i, u_i, v_{i+1}. This results in a rearrangement of the components of the unknown vector from the order $u_1, u_2, \cdots, u_R, v_1, v_2, \cdots, v_R$ in equations (4-1) and (4-2) to the order $v_1, u_1, v_2, u_2, \cdots, v_R, u_R$. The equations are then rearranged so that the first component equation of (4-1) is followed by the first component of (4-2). This is followed by the second component equation of (4-1) and then the second component of (4-2). This order is continued until the last component equation is the Rth component of (4-2). The resulting coefficient matrix is of the pentadiagonal form and takes the configuration shown at top of next page.

The diagonal terms for the component equations from (4-1) are the $b_i^{(2)}$, and those for the components of (4-2) are the $b_i^{(3)}$. Therefore, the equations from (4-1) have two terms to the right of the diagonal and one to the left, while those from (4-2) have two to the left and one to the right. Consequently, the matrix equation can be solved by the pentadiagonal algorithm. This

$$
\begin{bmatrix}
b_1^{(2)} & b_1^{(1)} & c_1^{(2)} & 0 & \cdots & & & & & & \\
b_1^{(4)} & b_1^{(3)} & c_1^{(4)} & 0 & \cdots & & & & & & \\
0 & a_2^{(1)} & b_2^{(2)} & b_2^{(1)} & c_2^{(2)} & 0 & \cdots & & & & \\
0 & a_2^{(3)} & b_2^{(4)} & b_2^{(3)} & c_2^{(4)} & 0 & \cdots & & & & \\
 & & \cdots & & & & & & & & \\
 & \cdots & 0 & a_i^{(1)} & b_i^{(2)} & b_i^{(1)} & c_i^{(2)} & 0 & \cdots & & \\
 & \cdots & 0 & a_i^{(3)} & b_i^{(4)} & b_i^{(3)} & c_i^{(4)} & 0 & \cdots & & \\
 & & \cdots & & & & & & & & \\
 & & & & & \cdots & 0 & a_R^{(1)} & b_R^{(2)} & b_R^{(1)} \\
 & & & & & \cdots & 0 & a_R^{(3)} & b_R^{(4)} & b_R^{(3)}
\end{bmatrix}
\begin{bmatrix} v_1 \\ u_1 \\ v_2 \\ u_2 \\ \cdots \\ v_i \\ u_i \\ \cdots \\ v_R \\ u_R \end{bmatrix}
=
\begin{bmatrix} d_1^{(1)} \\ d_1^{(2)} \\ d_2^{(1)} \\ d_2^{(2)} \\ \cdots \\ d_i^{(1)} \\ d_i^{(2)} \\ \cdots \\ d_R^{(1)} \\ d_R^{(2)} \end{bmatrix}
\tag{4-3}
$$

alternate method of solution of the finite difference equations for two-point boundary value, convective problems in two dependent variables may sometimes be more convenient to use than the bi-tridiagonal algorithm. Furthermore, the rearrangement procedure used above can be applied to determine equivalent forms of more complex systems of equations.

3. BAND FORM OF COMPLETE TRIDIAGONAL SYSTEMS

The complete bi-tridiagonal system of equations (3-10a) and (3-11a) contains two more coefficients in each component equation, these being $a_i^{(2)}$ and $c_i^{(1)}$ for (4-1) and $a_i^{(4)}$ and $c_i^{(3)}$ for (4-2). With the same two terms as diagonals, there now are three terms to the right of the diagonal in the components of (4-1) and three to the left in those of (4-2). As a result the complete bi-tridiagonal system of equations could be solved by an algorithm for a heptadiagonal matrix. Notice, however, that a complete heptadiagonal system cannot be solved by the bi-tridiagonal algorithm, for there are not three terms on each side of the diagonal in the component equations when the complete bi-tridiagonal equations are written in a band form.

In the previous chapter, it was observed that the finite difference equations for a number of physical problems result in a set of equations which have tridiagonal matrices for three dependent variables. The algorithm for the solution of this system is given in the Appendix. This system of equations can be written as three simultaneous matrix equations. They can be written as a single equation, with m taking the values successively of 1, 2, and 3 as shown below:

$$\begin{bmatrix} & \cdots & \\ a_i^{(3m-2)} & b_i^{(3m-2)} & c_i^{(3m-2)} \\ & \cdots & \end{bmatrix}\begin{bmatrix} \cdots \\ u_i \\ \cdots \end{bmatrix} + \begin{bmatrix} & \cdots & \\ a_i^{(3m-1)} & b_i^{(3m-1)} & c_i^{(3m-1)} \\ & \cdots & \end{bmatrix}\begin{bmatrix} \cdots \\ v_i \\ \cdots \end{bmatrix}$$

$$+ \begin{bmatrix} & \cdots & \\ a_i^{(3m)} & b_i^{(3m)} & c_i^{(3m)} \\ & \cdots & \end{bmatrix}\begin{bmatrix} \cdots \\ w_i \\ \cdots \end{bmatrix} = \begin{bmatrix} \cdots \\ d_i^{(m)} \\ \cdots \end{bmatrix} \qquad \text{(4-4a, b, c)}$$

These three equations can be written as one, and rearranged in the same manner as was the matrix for the bi-tridiagonal system above. The resulting matrix equation is

$$\begin{bmatrix} & & & & \cdots & & & & \\ a_i^{(1)} & a_i^{(2)} & a_i^{(3)} & b_i^{(1)} & b_i^{(2)} & b_i^{(3)} & c_i^{(1)} & c_i^{(2)} & c_i^{(3)} \\ a_i^{(4)} & a_i^{(5)} & a_i^{(6)} & b_i^{(4)} & b_i^{(5)} & b_i^{(6)} & c_i^{(4)} & c_i^{(5)} & c_i^{(6)} \\ a_i^{(7)} & a_i^{(8)} & a_i^{(9)} & b_i^{(7)} & b_i^{(8)} & b_i^{(9)} & c_i^{(7)} & c_i^{(8)} & c_i^{(9)} \\ & & & & \cdots & & & & \end{bmatrix}\begin{bmatrix} \cdots \\ u_i \\ v_i \\ w_i \\ \cdots \end{bmatrix} = \begin{bmatrix} \cdots \\ d_i^{(1)} \\ d_i^{(2)} \\ d_i^{(3)} \\ \cdots \end{bmatrix} \qquad \text{(4-4d)}$$

The diagonal terms for the three successive component equations are $b_i^{(1)}$, $b_i^{(5)}$, and $b_i^{(9)}$. For these three equations, there are five terms to the right of the diagonal and three to the left in the first equation, four on each side in the second equation, and three to the right and five to the left in the third component equation. Consequently, any system of equations which can be arranged into this form can be solved by the tri-tridiagonal algorithm. On the other hand, the system of equations given as (4-4d) can be solved by a solution algorithm for a band coefficient matrix with the eleven principal diagonals containing the only non-zero terms.

The general system of equations which is tridiagonal in four dependent variables becomes even more complex. There are now four equations of the form of (4-4a), each with four unknown vectors. These can be written as a single matrix equation and become as shown at top of next page.

The diagonal terms of this coefficient matrix are $b_i^{(1)}$, $b_i^{(6)}$, $b_i^{(11)}$, and $b_i^{(16)}$. There are a maximum of seven terms to each side of the diagonal in at least one of the equations. Therefore, this set of equations can be solved by the algorithm for a band matrix with the fifteen principal diagonals containing all the non-zero elements.

$$
\begin{bmatrix}
 & & & & & & \cdots & & & & & \\
a_i^{(1)} & a_i^{(2)} & a_i^{(3)} & a_i^{(4)} & b_i^{(1)} & b_i^{(2)} & b_i^{(3)} & b_i^{(4)} & c_i^{(1)} & c_i^{(2)} & c_i^{(3)} & c_i^{(4)} \\
a_i^{(5)} & a_i^{(6)} & a_i^{(7)} & a_i^{(8)} & b_i^{(5)} & b_i^{(6)} & b_i^{(7)} & b_i^{(8)} & c_i^{(5)} & c_i^{(6)} & c_i^{(7)} & c_i^{(8)} \\
a_i^{(9)} & a_i^{(10)} & a_i^{(11)} & a_i^{(12)} & b_i^{(9)} & b_i^{(10)} & b_i^{(11)} & b_i^{(12)} & c_i^{(9)} & c_i^{(10)} & c_i^{(11)} & c_i^{(12)} \\
a_i^{(13)} & a_i^{(14)} & a_i^{(15)} & a_i^{(16)} & b_i^{(13)} & b_i^{(14)} & b_i^{(15)} & b_i^{(16)} & c_i^{(13)} & c_i^{(14)} & c_i^{(15)} & c_i^{(16)} \\
 & & & & & & \cdots & & & & &
\end{bmatrix}
\times
\begin{bmatrix} \cdots \\ u_i \\ v_i \\ w_i \\ z_i \\ \cdots \end{bmatrix}
=
\begin{bmatrix} \cdots \\ d_i^{(1)} \\ d_i^{(2)} \\ d_i^{(3)} \\ d_i^{(4)} \\ \cdots \end{bmatrix}
\quad (4\text{-}5)
$$

4. GENERAL TRIDIAGONAL SET OF EQUATIONS

Most of the finite difference equations which have been developed for the numerical solution of parabolic and hyperbolic partial differential equations are, at most, tridiagonal in each of the dependent variables. It is convenient in developing a nomenclature for these equations to use one which utilizes this property of the equations. This nomenclature has been used in equations (3-10a) and (3-11a), as well as in the other equations discussed in this chapter. The general equation contains the dependent variables with three successive values of the subscript—namely, $i - 1$, i, and $i + 1$. All the coefficients for this equation are subscripted with i. The coefficients for a variable subscripted with i are called b; those for a variable subscripted with $i - 1$ are called a; and those for a variable subscripted with $i + 1$ are called c. The known term, which appears on the right side of the equation, is termed d. There must be as many differential equations to describe the physical problem as there are dependent variables. (It is assumed that dependent variables related only by an algebraic or transcendental equation need not be included in the simultaneous solution.) Consequently, the number of d vectors is equal to the number of differential equations, and the number of a, b, and c vectors is equal to the number of differential equations multiplied by the number of dependent variables. Since these two are equal, the number of a, b, and c vectors equals the square of the number of dependent variables.

For the nomenclature used herein, the various coefficient vectors are distinguished by superscripts. A generalization of this superscript nomenclature can be obtained for a general set of differential equations. Let S be the number of differential equations and, consequently, the number of dependent variables. Number the differential equations from 1 to S by the index L. Correspondingly, number the dependent variables from 1 to S by the index k. The dependent variables are then called $u^{(k)}$, where $k = 1, 2, \cdots, S$. The superscripts on the d vectors correspond directly to the index on the differential equation which gave rise to that set of finite difference equations. Thus, these vectors are called $d_i^{(L)}$, with $L = 1, 2, \cdots, S$.

A more complex superscript must be used for the a, b, and c vectors, as this superscript must assume integer values from 1 to S^2. The superscript used in all the examples above is $S(L - 1) + k$ for $u = u^{(1)}$, $v = u^{(2)}$, $w = u^{(3)}$, and $z = u^{(4)}$. The general set of finite difference equations then becomes

$$\sum_{k=1}^{S}(a_i^{[S(L-1)+k]}u_{i-1}^{(k)} + b_i^{[S(L-1)+k]}u_i^{(k)} + c_i^{[S(L-1)+k]}u_{i+1}^{(k)}) = d_i^{(L)} \qquad \text{for } 1 \leq L \leq S \quad (4\text{-}6)$$

For R finite difference equations corresponding to each differential equation, there will be R times S finite difference equations in the complete set. An alternate superscript notation can be used which might be easier to use when S is larger than 3 or 4. In this case, a double superscript can be used so that $a_i^{[S(L-1)+k]} = a_i^{(L,k)}$.

Either of these sets of nomenclature is convenient for use in storing the coefficients of the finite difference equations. The required number of coefficient vectors can be computed directly from the finite difference equations and put into storage under the system described above. For most physical problems a number of these vectors may be zero or constant. However, storage and programming inefficiences can be tolerated on problems of this type. A simple routine has been written to restore these coefficients in an arrangement which is suitable for solution by the general band algorithm. In performing this rearrangement, one should number the dependent variables so that the narrowest band results if some of the a and c vectors are all zero. For an example, see Section 2 of this chapter in which a special case of a bi-tridiagonal matrix was rearranged to a pentadiagonal band, whereas the complete bi-tridiagonal matrix forms a heptadiagonal band.

5. GENERAL BAND SET OF EQUATIONS

The algorithm for the solution of the general band matrix was developed by Peaceman (*8*), and his nomenclature for the coefficients will be used. In this form of the equations, all the non-zero coefficients are grouped as near the main diagonal as possible. In Peaceman's nomenclature, the general band equations are

$$A_j^{(M)}X_{j-M} + A_j^{(M-1)}X_{j-(M-1)} + \cdots + A_j^{(2)}X_{j-2} + A_j^{(1)}X_{j-1}$$
$$+ B_jX_j + C_j^{(1)}X_{j+1} + C_j^{(2)}X_{j+2} + \cdots + C_j^{(M-1)}X_{j+(M-1)}$$
$$+ C_j^{(M)}X_{j+M} = D_j \qquad \text{for } 1 \leq j \leq N \text{ with } N \geq M \qquad (4\text{-}7)$$

All the equations are written as one set, with a total of N equations in the set and with the subscript j indicating the position of the equation in the set. Each equation is assumed to have an equal number of terms on each side of the diagonal term, X_j. In the general equation, there are M terms on each side of the diagonal for a total of $2M + 1$ terms in each equation. The coefficient of the diagonal term, X_j, is called B; the coefficients of terms with subscripts less than j are called A; and those of terms with subscripts greater than j are called C. The known term on the right side of the equation is called D. All coefficients in the equation indexed by j are subscripted with j, and superscripts are used to distinguish among the various A and C vectors. As can be seen in equation (4-7), the superscript on each A or C is the number of terms it is to the left or right of the diagonal term. Peaceman has developed an algorithm for the solution of this set of equations which is convenient to use with this nomenclature. This algorithm is a generalization of the algorithms for the tridiagonal and pentadiagonal systems and is presented in the Appendix.

The formulation of the finite difference equations and the computation of the coefficients is usually made most easily in the general tridiagonal form. It is necessary, therefore, to obtain the correspondence between the two systems so that the coefficients can be computed in the general tridiagonal form and the equations can be solved in the general band form.

6. CONVERSION OF GENERAL TRIDIAGONAL FORM TO GENERAL BAND FORM

To obtain the narrowest band containing the fewest zeros within the band, which objective is desirable, one should follow the procedure used in the earlier sections of this chapter for 2, 3, and 4 tridiagonal systems. In the tridiagonal system there are S differential equations with R finite difference equations replacing each. Thus, the total number of equations in the band system, N, will equal R times S. In converting from one system to the other, both the order of the listing of the equations and the order of the components in the solution vector are changed. In the tridiagonal system there are S equations for each of the R values of the index i. The equations are reordered so that the S equations with the index $i = 1$ are the first S equations of the band system, with the index j assuming values from 1 to S. The next S equations for $(S + 1) \leq j \leq 2S$ are those for $i = 2$. This process is continued until the last S equations for $[S(R - 1) + 1] \leq j \leq RS$ are those for $i = R$. Within each set of S equations for a given value of the index i, the equations are arranged according to the index L, which designates the differential equation which gave rise to that particular finite difference equation.

If the unknowns of the tridiagonal system are considered as components of a single solution vector, they are arranged in the order $u_1^{(1)}, u_2^{(1)}, \cdots, u_R^{(1)}$, $u_1^{(2)}, u_2^{(2)}, \cdots, u_R^{(2)}, \cdots, u_1^{(S)}, u_2^{(S)}, \cdots, u_R^{(S)}$, where k is the index in the superscript which denotes the dependent variable, and $1 \leq k \leq S$. To obtain the narrowest band, these components are rearranged so that the S components with subscript $i = 1$ are the first S components for $1 \leq j \leq S$, the next S components for $(S + 1) \leq j \leq 2S$ are those for $i = 2$, and the last S components for $[S(R - 1) + 1] \leq j \leq RS$ are those for $i = R$. The components within a set for a constant value of i are ordered according to the index k.

For the rearrangement of both equations and components described above, the term on the main diagonal of each equation is the b from the tridiagonal set for $L = k$. Each equation contains $3S$ non-zero terms, but these are not arranged symmetrically about the diagonal in all equations. For $k = L = 1$, there are $(S - 1)$ b terms and S c terms to the right of the diagonal, and for $k = L = S$, there are $(S - 1)$ b terms and S a terms to the left of the diagonal; so $M = 2S - 1$. The total width of the band is $2M + 1$ or $4S - 1$. However, there are only $3S$ non-zero terms in each equation; so $S - 1$ terms in each equation of the band are zero. These zeros are located on the outside of the band, and they are distributed on either side of the band, depending on the value of L. For $L = 1$, all the $(S - 1)$ zeros are to the left of the diagonal, and for $L = S$, all zeros are to the right of the diagonal. The terms of the band which are zero are given in Table 4-1 with the conversion between the two systems.

The conversion of the coefficients in the tridiagonal system to those of the band system is given in Table 4-1. A FORTRAN program has been written to effect this rearrangement and to put the zeros in the proper places of the band system. An additional program has been written to relocate the solution to the $u_i^{(k)}$ vectors. In these programs, coefficients in the band system are indicated by double letters. To conform to FORTRAN language, S and R are called, respectively, IS and IR, and the superscripts are used as the second

Table 4-1. Conversion between Tridiagonal and Band Forms

Tridiagonal	*Band*	
$a_i^{[S(L-1)+k]}$	$A_{S(i-1)+L}^{(S+L-k)}$	
$b_i^{[S(L-1)+k]}$	$A_{S(i-1)+L}^{(L-k)}$	$k < L$
$b_i^{[S(L-1)+k]}$	$B_{S(i-1)+L}$	$k = L$
$b_i^{[S(L-1)+k]}$	$C_{S(1-1)+L}^{(k-L)}$	$k > L$
$c_i^{[S(L-1)+k]}$	$C_{S(i-1)+L}^{(S-L+k)}$	
$d_i^{(L)}$	$D_{S(i-1)+L}$	
$u_i^{(k)}$	$X_{S(i-1)+k}$	
0	$A_j^{(P)}$ $(S + L) \leq P \leq (2S - 1)$;	$L < S$
0	$C_j^{(P)}$ $(2S - L + 1) \leq P \leq (2S - 1)$;	$L > 1$

subscript of the double subscripted variables. This program is presented in the Appendix. Finite difference equations which contain the dependent variables at no more than three adjacent points can be solved for any number of dependent variables with this program and the general band algorithm.

Chapter Five

NONLINEAR PARABOLIC EQUATIONS

I. INTRODUCTION

One of the most important applications of numerical methods of solution is to nonlinear, partial differential equations. Several methods for solving quasi-linear equations have been developed that result in linear finite difference equations which can be solved by existing algorithms and which do not involve excessive iteration. These methods are used with finite difference analogs which are centered midway in time between the old and the new time steps. They are explained most readily when applied to parabolic equations; consequently, these equations will be studied first.

The general quasi-linear parabolic equation is

$$[a(u)]\frac{\partial^2 u}{\partial x^2} + [b(u)]\frac{\partial u}{\partial x} + [c(u)]u = \frac{\partial u}{\partial t} \tag{5-1}$$

In order that the equation be quasi-linear, these coefficients must be functions of u only and not of its derivatives. A discussion of methods for handling a few other types of nonlinear equations is included in a later section. The general finite difference equation to be used for solving (5-1) is the Crank-Nicolson equation, and various methods of handling the nonlinear coefficients are used in conjunction with this method. For many quasi-linear equations, the boundary conditions are linear and can be handled by methods discussed previously. Consequently, no treatment of the boundary equations will be included in this discussion. A later chapter is devoted to methods of handling nonlinear boundary conditions.

In a discussion of the various methods for solving numerically quasi-linear parabolic equations, it it sufficient to consider the simplest equation of this type. The only new technique is the method of handling the nonlinear coefficients; and, since all nonlinear coefficients are handled similarly, it is sufficient that the equation contain only one such coefficient. The following discussion, therefore, will be concerned with

$$[a(u)]\frac{\partial^2 u}{\partial x^2} = \frac{\partial u}{\partial t} \tag{5-1a}$$

2. ITERATION USING OLD VALUE

The Crank-Nicolson analogs to the derivatives are centered about the time level $t_{n+1/2}$. An analog to the nonlinear coefficient, $a(u)$, is required at this time level; and, if the resulting finite difference equations are to be linear, this analog must not contain values of u at the time level t_{n+1}. The simplest such analog is obtained by evaluating $a(u)$ at the old time level and using $a(u_n)$ for $a(u_{n+1/2})$. If the function $a(u)$ does not change very rapidly with u, the solution to the resulting finite difference equations should be fairly near the correct values. These values can be improved by next evaluating $a(u_{n+1/2})$ as $a[(u_n + u_{n+1}^{(1)})/2]$, where $u_{n+1}^{(1)}$ is the value obtained when $a(u_n)$ was used for $a(u_{n+1/2})$. The result of a continuation of this procedure is the following iterative equations:

$$\left[a\left(\frac{u_{i,n} + u_{i,n+1}^{(m)}}{2}\right)\right]\frac{1}{2}\,\Delta_x^2(u_{i,n} + u_{i,n+1}^{(m+1)}) = \frac{u_{i,n+1}^{(m+1)} - u_{i,n}}{\Delta t} \tag{5-2}$$

where

$$\Delta_x^2 u_{i,n} = \frac{u_{i+1,n} - 2u_{i,n} + u_{i-1,n}}{(\Delta x)^2}$$

and

$$u_{i,n+1}^{(0)} = u_{i,n}$$

Iteration is continued until $u_{i,n+1}^{(m+1)} = u_{i,n+1}^{(m)}$ within a predetermined tolerance.

The resulting finite difference equations are linear in $u_{i,n+1}^{(m+1)}$ with the coefficient matrix being tridiagonal so that the Thomas algorithm can be used for the solution. This method should converge in three or four iterations. Notice, however, that the first analog to $a(u_{i,n+1/2})$ is completely forward, so that there will be some limitations on the size of the time increment to ensure stability.

3. FORWARD PROJECTION OF COEFFICIENT TO HALF LEVEL IN TIME

Douglas (*9*) has devised a method for projecting the value of u_i to the half-time level for use in the nonlinear coefficients. This method has less stringent restrictions on the time-step size for stability and converges more rapidly than the method described above. Essentially, this method results in a two-step iteration process. The restrictions for stability are rather complex; reference to the original paper is recommended for a discussion of these restrictions.

For this method the value of the dependent variable at the half level in time is obtained from a truncated Taylor series as follows:

$$u_{i,n+\frac{1}{2}} = u_{i,n} + \left(\frac{\partial u}{\partial t}\right)_{i,n}\left(\frac{\Delta t}{2}\right) + \left(\frac{\partial^2 u}{\partial t^2}\right)_{i,n}\frac{1}{2!}\left(\frac{\Delta t}{2}\right)^2 + \cdots + \tag{5-3}$$

The series in equation (5-3) is truncated after the second term to obtain a second-order-correct analog to $u_{i,n+\frac{1}{2}}$. The time derivative in this analog is then obtained from equation (5-1a). The resulting finite difference analog for $u_{i,n+\frac{1}{2}}$ to be used in the nonlinear coefficient is

$$u_{i,n+\frac{1}{2}} = u_{i,n} + \frac{\Delta t}{2}\,[a(u_{i,n})]\,\Delta_x^2 u_{i,n} \tag{5-3a}$$

This value is then used in evaluating $a(u)$ for use in the Crank-Nicolson analog to (5-1a). The resulting finite difference equation can be written as

$$\left\{a\left[u_{i,n} + \frac{\Delta t}{2}\,a(u_{i,n})\,\Delta_x^2 u_{i,n}\right]\right\}\frac{1}{2}\,\Delta_x^2(u_{i,n} + u_{i,n+1}) = \frac{u_{i,n+1} - u_{i,n}}{\Delta t} \tag{5-4}$$

The values of $u_{i,n+1}$ which result from the application of (5-4) can be corrected by an iteration procedure similar to that described in the previous section. The need for such a correction can be determined experimentally and usually is found unnecessary. It might be more advisable to decrease the time step rather than to iterate.

The actual application of (5-4) is performed in two steps. First the values of $u_{i,n+\frac{1}{2}}$ are determined explicitly from (5-3a), and these values are used to evaluate $a(u_{i,n+\frac{1}{2}})$. These values are then used to compute the elements of the coefficient matrix for (5-4), and this is solved by the Thomas algorithm. Such a procedure has been proved to be very efficient for the numerical solution of a number of quasi-linear, partial differential equations.

4. BACKWARD TAYLOR SERIES PROJECTION

An analog to $u_{i,n+\frac{1}{2}}$ for use in the nonlinear coefficients can also be obtained from a truncated Taylor series written about the time level $t_{n+\frac{1}{2}}$. This analog has even fewer elements of a forward difference than that defined in (5-3a), although the computation of $u_{i,n+\frac{1}{2}}$ may be slightly more time-consuming than that by the previous method. In this case, the Taylor series is

$$u_{i,n} = u_{i,n+\frac{1}{2}} - \left(\frac{\partial u}{\partial t}\right)_{i,n+\frac{1}{2}}\frac{\Delta t}{2} + \left(\frac{\partial^2 u}{\partial t^2}\right)_{i,n+\frac{1}{2}}\frac{1}{2!}\left(\frac{\Delta t}{2}\right)^2 - \cdots - \tag{5-5}$$

Again, the Taylor series is truncated after the second term to obtain a second-order-correct analog, and the time derivative is obtained from (5-1a). In this case, however, the time derivative should be evaluated at the time level $t_{n+1/2}$; but, if this were done for the complete analog, the resulting equations would be nonlinear. Consequently the space derivative is evaluated at $t_{n+1/2}$, but the nonlinear coefficient is evaluated at t_n. The resulting finite difference equations can be written as

$$[a(u_{i,n})]\,\Delta_x^2 u_{i,n+1/2} = \frac{u_{i,n+1/2} - u_{i,n}}{\Delta t/2} \tag{5-5a}$$

This equation is not explicit as is (5-3a), but the resulting coefficient matrix is tridiagonal. The values of $u_{i,n+1/2}$ can thus be readily obtained, and these are used in a Crank-Nicolson analog to (5-1a), which is

$$[a(u_{i,n+1/2})]\,\frac{1}{2}\,\Delta_x^2(u_{i,n} + u_{i,n+1}) = \frac{u_{i,n+1} - u_{i,n}}{\Delta t} \tag{5-6}$$

The values obtained from the two-step process of using (5-5a) and then (5-6) can also be improved by an iteration technique similar to that described in Section 2. It is obvious that a centered or Crank-Nicolson type of evaluation of $u_{i,n+1/2}$ would be more convenient to use with (5-6) than is (5-5a). However, the projection of equation (5-5a) can be used most conveniently when the backward equation is used for equation (5-1a).

5. CENTERED TAYLOR SERIES PROJECTION

The centered Taylor series projection for $u_{i,n+1/2}$ can be obtained from the Taylor series for $u_{i,n+1/2}$ and $u_{i,n}$ written about the level $t_{n+1/2}$. In this case $a(u_{i,n})$ is used for $a(u_{i,n+1/2})$, and the resulting finite difference equation is

$$[a(u_{i,n})]\,\frac{1}{2}\,\Delta_x^2(u_{i,n} + u_{i,n+1/2}) = \frac{u_{i,n+1/2} - u_{i,n}}{\Delta t/2} \tag{5-7}$$

The values of $u_{i,n+1/2}$ obtained from this implicit equation are then used in (5-6) to obtain the values of $u_{i,n+1}$. The basic difference between this method and that described in Section 4 is that $a(u_{i,n})$ is used for $a(u_{i,n+1/4})$ in this method and for $a(u_{i,n+1/2})$ in the method of Section 4. The use of (5-7) and (5-6) successively should prove easy to program, since the solution of (5-6) involves only an adjustment of some of the cofficients of (5-7) which are based on the solution of (5-7). In fact, the solution to (5-7) can be considered as the first iterative values of (5-6) rather than as values of the dependent variable at the half time level. This method is called a predictor-corrector method by Douglas.

6. APPLICATION TO NONLINEAR ORDINARY EQUATIONS WITH SPLIT BOUNDARY CONDITIONS

The classical methods for solving ordinary differential equations, such as the Runge-Kutta method, result in iterative solutions for split boundary conditions. For linear equations with split boundary conditions, the method of Chapter One results in a direct solution, but for nonlinear equations an iterative solution can be attained by an adaptation of the methods discussed above. Thus an alternate iterative method of solution is afforded.

This method of solving ordinary equations is one in which each stage of the iteration may be regarded as a time step of an unsteady-state problem, with the starting values, or first guess, being regarded as the initial conditions. The time increment, Δt, of the unsteady-state problem may be regarded as the iteration parameter, ϵ, of the steady-state problem. Actually, any of the methods described above may be adapted for the solution of the ordinary differential equation. Several of these methods will contain two or more steps in one complete iteration.

Generally, the iteration parameter, ϵ, can be increased in size as the solution is approached. Furthermore, since the intermediate values are of no interest, the initial iteration parameter can be larger than the initial time step which is used in an unsteady-state problem. In fact, it can be shown that the Crank-Nicolson equation for linear differential equations will converge to the correct steady-state solution even when early values of the time increment are so large that there is considerable truncation error in values of u at early time levels. The only consideration of interest in the solution of a nonlinear, ordinary differential equation is that convergence to the correct steady-state solution be attained in the least number of iterations. It is desirable therefore, to use the sequence of values for the iteration parameter which will result in such convergence. Peaceman and Rachford (*10*) have found such a sequence for the solution of elliptic equations which describe steady-state conduction problems in two or more dimensions. For problems in one space dimension, the solution can be obtained very rapidly on modern high-speed computers with any reasonable value of the parameter.

7. SOLUTION OF SIMULTANEOUS EQUATIONS

Any of the several methods discussed above can be used for the solution of nonlinear, coupled partial differential equations. The extension of these methods to coupled equations is straightforward if the first time derivative of each of the variables appears in one equation which does not contain any other time derivatives. In some cases, a rearrangement of the equations may be necessary to put them into this form. An early application of these methods to coupled equations was reported by von Rosenberg *et al.* (*11*). The equations solved were the enthalpy and reactant material balances describing a packed bed reactor and were coupled through the term describing the chemical reaction. The method of Section 3 was used for the nonlinear term, and this method proved to be satisfactory with no further iteration. Harris and Weaver (*12*) used this same method to solve the four simultaneous equations which describe the transmission of a potential impulse along a nerve. Three of these equations contained only time derivatives, while the fourth contained a time derivative and the second space derivative. The method of Section 3 was used; and, since some of the time derivatives were described by highly nonlinear functions, a check was made to determine if further iteration improved the solution. Such iteration was found to be unnecessary.

8. METHODS FOR OTHER TYPES OF NONLINEARITIES

While a large fraction of the problems encountered by an engineer can be described by quasi-linear equations, there are a significant number of problems in which other types of nonlinearities arise. Douglas and Jones (*13*) have proposed an adaptation of the methods described above which will apply to another class of nonlinear equations. This class of equations is defined by

$$\frac{\partial^2 u}{\partial x^2} = \left[g_1\left(x, t, u, \frac{\partial u}{\partial x}\right) \right] \frac{\partial u}{\partial t} + g_2\left(x, t, u, \frac{\partial u}{\partial x}\right) \tag{5-8}$$

To solve this equation numerically, one must first obtain $u_{i,n+1/2}$ for all values of i. These values may be obtained by any of the three projection methods described above. The actual finite difference equations are then written according to the Crank-Nicolson technique with

$$g_1 = g_1\left[x_i, t_{n+1/2}, u_{i,n+1/2}, \left(\frac{u_{i+1,n+1/2} - u_{i-1,n+1/2}}{2(\Delta x)} \right) \right] \tag{5-9}$$

and with g_2 evaluated similarly. This method of evaluating g_1 and g_2 for use in the Crank-Nicolson equation results in linear finite difference equations which can be solved by one of the existing algorithms. Furthermore, Douglas and Jones have shown that this method preserves the unconditional stability of the Crank-Nicolson equations, though the time analog has a truncation error which is of the order of $(\Delta t)^{3/2}$.

This method can be used for any type of nonlinearity for which the coefficients of the derivatives are functions of the dependent variable, u, and of its first space derivative, $(\partial u/\partial x)$. One example of this type of nonlinearity arises in cases for which the transport coefficient, such as the thermal conductivity, is a function of the dependent variable. Problems of this type account for most of the nonlinear equations, other than quasi-linear ones, which are encountered by engineers. The simplest differential equation of this type is

$$[k(u)]\frac{\partial^2 u}{\partial x^2} + [k'(u)]\left(\frac{\partial u}{\partial x}\right)^2 = \frac{\partial u}{\partial t} \tag{5-10}$$

where $k'(u) = (dk/du)$. For any given problem, the conductivity, k, will be a known function of the dependent variable u; thus, its derivative will also be a known function. By use of the method described above, the following finite difference analog is obtained:

$$[k(u_{i,n+\frac{1}{2}})]\,\frac{1}{2}\,\Delta_x^2(u_{i,n+1} + u_{i,n}) + [k'(u_{i,n+\frac{1}{2}})]\,\delta_x u_{i,n+\frac{1}{2}} \times \left[\frac{1}{2}\,\delta_x(u_{i,n} + u_{i,n+1})\right] = \frac{u_{i,n+1} - u_{i,n}}{\Delta t} \tag{5-11}$$

where

$$\delta_x u_{i,n} = \frac{u_{i+1,n} - u_{i-1,n}}{2(\Delta x)}$$

The values for $u_{i,n+\frac{1}{2}}$ can be obtained from any of the three projection methods. With these values used in (5-11), the resulting finite difference equations are linear, and they can be solved by the Thomas algorithm.

It is informative to examine the form of the coefficients of the three unknowns which appear in (5-11). Equation (5-11) can be rearranged to

$$\frac{1}{(\Delta x)^2}\left[k_{i,n+\frac{1}{2}} + \frac{1}{2}\,k'_{i,n+\frac{1}{2}}\left(\frac{u_{i+1,n+\frac{1}{2}} - u_{i-1,n+\frac{1}{2}}}{2}\right)\right][u_{i+1,n+1} + u_{i+1,n}]$$
$$+ \frac{1}{(\Delta x)^2}\left[k_{i,n+\frac{1}{2}} - \frac{1}{2}\,k'_{i,n+\frac{1}{2}}\left(\frac{u_{i+1,n+\frac{1}{2}} - u_{i-1,n+\frac{1}{2}}}{2}\right)\right][u_{i-1,n+1} + u_{i-1,n}]$$
$$+ \left[-\frac{2k_{i,n+\frac{1}{2}}}{(\Delta x)^2} - \frac{2}{\Delta t}\right]u_{i,n+1} + \left[-\frac{2k_{i,n+\frac{1}{2}}}{(\Delta x)^2} + \frac{2}{\Delta t}\right]u_{i,n} = 0 \tag{5-11a}$$

In this relation, $k_{i,n+\frac{1}{2}} = k(u_{i,n+\frac{1}{2}})$ and $k'_{i,n+\frac{1}{2}} = k'(u_{i,n+\frac{1}{2}})$. The coefficient of $u_{i+1,n+1}$ is a truncated Taylor series for $k(u_{i+\frac{1}{2},n+\frac{1}{2}})$. This relation can be obtained from

$$k_{i+\frac{1}{2},n+\frac{1}{2}} = k_{i,n+\frac{1}{2}} + k'_{i,n+\frac{1}{2}}\left(\frac{\partial u}{\partial x}\right)\frac{\Delta x}{2} + \cdots + \tag{5-12}$$

When the space derivative, $(\partial u/\partial x)$, in (5-12) is replaced by the finite difference analog of (1-8), the coefficient of $u_{i+1,n+1}$ in (5-11a) is shown to be a second-order-correct analog of $k_{i+\frac{1}{2},n+\frac{1}{2}}$. In a similar manner, it can be shown that the coefficient of $u_{i-1,n+1}$ is a second-order-correct analog for $k_{i-\frac{1}{2},n+\frac{1}{2}}$. In the coefficient of $u_{i,n+1}$, the contribution from the second space derivative, $2k_{i,n+\frac{1}{2}}$, may be interpreted as the sum of $k_{i+\frac{1}{2},n+\frac{1}{2}}$ and $k_{i-\frac{1}{2},n+\frac{1}{2}}$. This interpretation of these coefficients indicates another method of formulating a finite difference analog to (5-10). This alternate method is actually a better analog for the representation of the conductivity, k; furthermore, it is much more convenient to use in a numerical solution, especially when k is a function of several dependent variables.

The formulation of this alternate equation is most easily developed from the compact form of the original differential equation. This form is

$$\frac{\partial}{\partial x}\left[k(u)\frac{\partial u}{\partial x}\right] = \frac{\partial u}{\partial t} \tag{5-10a}$$

The finite difference analog to the space derivative in (5-10a) is formulated in the manner discussed in Section 8 of Chapter One for the second space derivative. In this method, the first space derivative is approximated at $x_{i+\frac{1}{2}}$ and $x_{i-\frac{1}{2}}$ by the analogs over one space increment. The conductivity must, therefore, be evaluated at these same points. These evaluations are

$$k_{i+\frac{1}{2},n+\frac{1}{2}} = k\left(\frac{u_{i+1,n+\frac{1}{2}} + u_{i,n+\frac{1}{2}}}{2}\right) \tag{5-13a}$$

and

$$k_{i-\frac{1}{2},n+\frac{1}{2}} = k\left(\frac{u_{i-1,n+\frac{1}{2}} + u_{i,n+\frac{1}{2}}}{2}\right) \tag{5-13b}$$

The analog to (5-10a) then becomes

$$\frac{k_{i+\frac{1}{2},n+\frac{1}{2}}\frac{1}{2}\delta_x(u_{i+\frac{1}{2},n+1} + u_{i+\frac{1}{2},n}) - k_{i-\frac{1}{2},n+\frac{1}{2}}\frac{1}{2}\delta_x(u_{i-\frac{1}{2},n+1} + u_{i-\frac{1}{2},n})}{\Delta x} = \frac{u_{i,n+1} - u_{i,n}}{\Delta t} \tag{5-14}$$

where

$$\delta_x u_{i+\frac{1}{2},n} = \frac{u_{i+1,n} - u_{i,n}}{\Delta x}$$

These finite difference equations are linear and can be solved by one of the available algorithms.

In this case, it is also informative to examine the form of the coefficients in this equation. These coefficients can be obtained from

$$\frac{k_{i+1/2,n+1/2}}{(\Delta x)^2}(u_{i+1,n+1} + u_{i+1,n}) + \frac{k_{i-1/2,n+1/2}}{(\Delta x)^2}(u_{i-1,n+1} + u_{i-1,n})$$
$$- \left[\frac{k_{i+1/2,n+1/2} + k_{i-1/2,n+1/2}}{(\Delta x)^2} + \frac{2}{\Delta t}\right]u_{i,n+1}$$
$$+ \left[\frac{2}{\Delta t} - \frac{k_{i+1/2,n+1/2} + k_{i-1/2,n+1/2}}{(\Delta x)^2}\right]u_{i,n} = 0 \quad \text{(5-14a)}$$

A comparison of the coefficients of (5-14a) with those in (5-11a) will indicate that the interpretation of the coefficients in (5-11a) given above is a correct one. However, the approximations of these values in (5-14a) are generally better ones than the approximations in (5-11a). A comparison of these analogs is made in the next section.

The evaluation of the coefficients in (5-14a) should also be easier than the evaluation of those in (5-11a). There is no need in (5-14a) to evaluate the derivative of the conductivity or to evaluate an analog of the first space derivative of u at $t_{n+1/2}$. Equation (5-14a) has even more advantages over (5-11a) when k is a function of two or more dependent variables. In this case, the expanded form of the differential equation, comparable to (5-10), contains one or more additional nonlinear terms similar to the second term on the left side of (5-10). In the compact form of (5-10a), however, the equation is almost identical to (5-10a); and the resulting finite difference equation is only little more trouble to evaluate than is (5-14).

In some empirical correlations, the eddy transport coefficients are expressed as functions of the spatial derivative of one of the dependent variables. The form of (5-14) is readily applicable to this type of function also. The coefficient $k_{i+1/2,n+1/2}$ must be evaluated as a function of $(\partial u/\partial x)_{i+1/2,n+1/2}$, for example. A second-order-correct analog of this derivative can be computed from the values of the dependent variable, u, at the $t_{n+1/2}$ time level. An example is

$$\left(\frac{\partial u}{\partial x}\right)_{i+1/2,n+1/2} = \frac{u_{i+1,n+1/2} - u_{i,n+1/2}}{(\Delta x)} \quad \text{(5-15)}$$

The values of the transport coefficient can thus be readily computed for use in equation (5-14) even when these coefficients are functions of spatial derivatives of dependent variables.

9. COMPARISON OF METHODS FOR COMPUTING FUNCTIONS OF DEPENDENT VARIABLES

The discrete variables used in finite difference equations assume values only at discrete points separated by finite differences. In some of the methods described above for the solution of nonlinear differential equations, it is necessary to evaluate a function of a dependent variable at a point between two of these discrete points. In some cases it is necessary to evaluate $f(u)$ at $x_{i-½}$ when u is known at x_i and at x_{i-1}. The evaluation was always made as

$$f_{i-½} = f\left(\frac{u_i + u_{i-1}}{2}\right) \tag{5-16}$$

in the above discussion. This evaluation could also be made as

$$f_{i-½} = \frac{f(u_i) + f(u_{i-1})}{2} \tag{5-17}$$

The evaluation by (5-16) makes use of a second-order-correct analog to evaluate u as a function of x and of the exact functional relationship between f and u. The evaluation by (5-17), on the other hand, makes use of a second-order-correct analog to evaluate f as a function of x. Equation (5-16) is usually preferred, since the size of the length increments must be chosen to control the truncation error in the relationship between u and x. This is the only truncation error incurred by the use of (5-16). There are cases, however, in which (5-17) will give a more exact value than (5-16).

A similar situation is found when it is necessary to evaluate $f(u)$ at $t_{n+½}$ when u is not known at any time level greater than t_n. In all the projection methods discussed above, the dependent variable, u, was first evaluated at time level $t_{n+½}$ by a truncated Taylor series. The value of $f(u)$ was then determined with this value for $u_{n+½}$ from the exact relationship between f and u. The value of $f_{n+½}$ can also be determined from a Taylor series for the function f, thus:

$$f_{n+½} = f_n + \left(\frac{df}{du}\right)_n \left(\frac{\partial u}{\partial t}\right)_n \frac{\Delta t}{2} + \cdots + \tag{5-18}$$

This method is almost identical to that used in equation (5-12) for determining the value of $k_{i+½,n+½}$. The two methods are directly comparable to the two methods defined in (5-16) and (5-17). As in that case, it should usually be better to determine the value of the dependent variable with a second-order-correct analog and use this value in the exact function than to use a second-order-correct analog in the functional relationship between the function, $f(u)$, and the independent variable. All the projection methods described above make use of the former method.

EXAMPLE 5-1. ISOTHERMAL FLOW REACTOR WITH SECOND ORDER REACTION

$$\frac{\partial^2 u}{\partial x^2} - s\frac{\partial u}{\partial x} - ru^2 = \frac{\partial u}{\partial t}$$

$$u(x, 0) = 0 \qquad \text{all } x$$

$$\frac{\partial u}{\partial x} = 0 \qquad \text{at } x = 1, \text{ all } t$$

$$\frac{\partial u}{\partial x} + s(1 - u) = 0 \qquad \text{at } x = 0, \text{ all } t$$

Define $x_i = (i - \tfrac{1}{2})\,\Delta x$. Use the Crank-Nicolson method.
Use the forward projection for $u_{i,n+\frac{1}{2}}$.
Program for R increments, so $\Delta x = 1/R$.
The finite difference equations are:
For $2 \leq i \leq (R - 1)$:

$$\left[1 + \frac{s(\Delta x)}{2}\right]u_{i-1,n+1} + \left[-2 - r(\Delta x)^2 u_{i,n+\frac{1}{2}} - \frac{2(\Delta x)^2}{\Delta t}\right]u_{i,n+1}$$
$$+ \left[1 - \frac{s(\Delta x)}{2}\right]u_{i+1,n+1} = -\left[1 + \frac{s(\Delta x)}{2}\right]u_{i-1,n} - \left[1 - \frac{s(\Delta x)}{2}\right]u_{i+1,n}$$
$$+ \left[2 + r(\Delta x)^2 u_{i,n+\frac{1}{2}} - \frac{2(\Delta x)^2}{\Delta t}\right]u_{i,n}$$

with

$$u_{i,n+\frac{1}{2}} = u_{i,n} + \frac{\Delta t}{2(\Delta x)^2}\left\{\left[1 + \frac{s(\Delta x)}{2}\right]u_{i-1,n} + \left[1 - \frac{s(\Delta x)}{2}\right]u_{i+1,n} - [2 + r(\Delta x)^2 u_{i,n}]u_{i,n}\right\}$$

For $i = 1$:

$$\left\{-r(\Delta x)^2 u_{1,n+\frac{1}{2}} - \frac{2(\Delta x)^2}{\Delta t} - \left[1 + \frac{s(\Delta x)}{2}\right]\right\}u_{1,n+1} + \left[1 - \frac{s(\Delta x)}{2}\right]u_{2,n+1}$$
$$= -\left[1 - \frac{s(\Delta x)}{2}\right]u_{2,n} + \left\{r(\Delta x)^2 u_{1,n+\frac{1}{2}} - \frac{2(\Delta x)^2}{\Delta t} + \left[1 + \frac{s(\Delta x)}{2}\right]\right\}u_{1,n} - 2s(\Delta x)$$

with

$$u_{1,n+\frac{1}{2}} = u_{1,n} + \frac{\Delta t}{2(\Delta x)^2}\left\{\left[1 - \frac{s(\Delta x)}{2}\right]u_{2,n} + s(\Delta x) - \left[1 + \frac{s(\Delta x)}{2} + r(\Delta x)^2 u_{1,n}\right]u_{1,n}\right\}$$

For $i = R$:

$$\left[1 + \frac{s(\Delta x)}{2}\right]u_{R-1,n} + \left\{-r(\Delta x)^2 u_{R,n+\frac{1}{2}} - \frac{2(\Delta x)^2}{\Delta t} - \left[1 + \frac{s(\Delta x)}{2}\right]\right\}u_{R,n+1}$$

$$= -\left[1 + \frac{s(\Delta x)}{2}\right]u_{R-1,n} + \left\{r(\Delta x)^2 u_{R,n+\frac{1}{2}} - \frac{2(\Delta x)^2}{\Delta t} + \left[1 + \frac{s(\Delta x)}{2}\right]\right\}u_{R,n+1}$$

with

$$u_{R,n+\frac{1}{2}} = u_{R,n} + \frac{\Delta t}{2(\Delta x)^2}\left\{\left[1 + \frac{s(\Delta x)}{2}\right]u_{R-1,n} - \left[1 + \frac{s(\Delta x)}{2} + r(\Delta x)^2 u_{R,n}\right]u_{R,n}\right\}$$

Define

$$A = 1 + \frac{s(\Delta x)}{2}; \qquad C = 1 - \frac{s(\Delta x)}{2}$$

Call $u_{i,n+\frac{1}{2}} = UH$.
Use the Thomas algorithm:
Call $\beta_i = BT(I)$ and $\gamma_i = G(I)$
Call the total number of increments, IR.
See Example 2-2 for values of MM.

This program (on next page) was run for $IR = 20$, $s = 10$, $R = 5$. Steady state was reached after 39 time steps. With WATFOR compiler on an IBM 7044, the running time was 25 seconds for 39 time steps.

```
      DIMENSION U(100), BT(100), G(100)
      READ (5,30) R,S,IR,MM,NT
   30  FORMAT(2E10.4,3I3)
      DO 1 I = 1,IR
    1 U(I) = 0.
      IRM1=IR-1
      RR=IR
      DX=1./RR
      SDX=S*DX
      TSDX=2.*SDX
      SDXT=SDX/2.
      A=1.+SDXT
      C=1.-SDXT
      RDX=R*DX*DX
      T=0.
      DT=DX*DX
      N=0
      M=1
      WRITE (6,20)
      WRITE (6,21) N,T,DT
      WRITE (6,22) (U(I), I=1,IR)
      WRITE (6,23)
   20  FORMAT (1H1)
   21  FORMAT (1H0, I5, 2F11.4)
   22  FORMAT (1H0, (10F11.4))
   23  FORMAT (1H )
      DO 2 N=1,NT
      T=T+DT
      M=M+1
      DXT=(2.*DX*DX)/DT
      UH=U(1)+(C*U(2)-(A+RDX*U(1))*U(1)+SDX)/DXT
      BT(1) = -RDX*UH-DXT-A
      D= -C*U(2) + (RDX*UH -DXT+A)*U(1)-TSDX
      G(1)=D/BT(1)
      DO 3 I=2,IRM1
      UH=U(I)+(A*U(I-1)+C*U(I+1)-(2.+RDX*U(I))*U(I))/DXT
      B=-2.-RDX*UH-DXT
      BT(I)=B-A*C/BT(I-1)
      D=-A*U(I-1) -C*U(I+1) +(2.+RDX*UH-DXT) * U(I)
    3 G(I)=(D-A*G(I-1))/BT(I)
      UH=U(IR) + (A*U(IRM1)-(A+RDX*U(IR))*U(IR))/DXT
      B=-RDX*UH-DXT-A
      BT(IR)=B-A*C/BT(IRM1)
      D=-A*U(IRM1) + (RDX*UH-DXT+A) *U(IR)
      U(IR)=(D-A*G(IRM1))/BT(IR)
      DO 4 J=1,IRM1
      I=IR-J
    4 U(I)=G(I) - C*U(I+1)/BT(I)
      WRITE (6,21) N,T,DT
      WRITE (6,22) (U(I), I=1,IR)
      WRITE (6,23)
      IF (M-MM) 2,5,5
    5  WRITE (6,20)
      M = 0
    2  DT = 1.1 * DT
      CALL EXIT
      END
```

Chapter Six

NONLINEAR HYPERBOLIC EQUATIONS

1. ITERATION FOR HYPERBOLIC EQUATIONS

Most of the methods described in the previous chapter can be adapted for use with quasi-linear, hyperbolic equations. As one of the more important applications of numerical methods of solution of hyperbolic equations is to those with split boundary conditions, thc methods will be illustrated on this system. Consider the equations

$$-b_1\left(\frac{\partial u}{\partial x}\right) - [c_1(u, v)](u - v) = \frac{\partial u}{\partial t} \tag{6-1}$$

$$+b_2\left(\frac{\partial v}{\partial x}\right) + [c_2(u, v)](u - v) = \frac{\partial v}{\partial t} \tag{6-2}$$

subject to the boundary conditions of equation (3-9). More generally, the coefficients of the space derivatives could also be functions of u and v, but the methods of handling the nonlinear terms would be the same, and the illustration would be more cumbersome.

The nomenclature and indexing for the centered difference method applied to a split boundary value problem is illustrated in Figure 3-2. The derivatives are centered about the point $x_{i-\frac{1}{2}}$, $t_{n+\frac{1}{2}}$ for the variable u and about the point $x_{i+\frac{1}{2}}$, $t_{n+\frac{1}{2}}$ for the variable v. Actually, this is the same point in space, but the x index is shifted for the two variables. It is necessary then to evaluate $u_{i-\frac{1}{2},n+\frac{1}{2}}$ and $v_{i+\frac{1}{2},n+\frac{1}{2}}$ for determining $c_1(u, v)$ and $c_2(u, v)$ in the centered difference equation.

A method of straight iteration using old values similar to that described in Section 2 of the previous chapter can be used. The average of the two values at the level t_n must be used for the starting value; as a result, the starting value of the nonlinear coefficient $c_1(u, v)$ is

$$c_1^{(0)}(u_{i-\frac{1}{2},n+\frac{1}{2}}, v_{i+\frac{1}{2},n+\frac{1}{2}}) = c_1[\tfrac{1}{2}(u_{i-1,n} + u_{i,n}), \tfrac{1}{2}(v_{i,n} + v_{i+1,n})]$$

The starting value for $c_2(u, v)$ is defined similarly. With c_1 and c_2 being known, the centered difference equations, (3-10) and (3-11), can be used to compute the values of $u_{i,n+1}$ and $v_{i,n+1}$ by one of the algorithms. These values can then be improved by an iterative process similar to that described by (5-2).

2. FORWARD PROJECTION

The forward projection method described in Section 3 of Chapter Five is also readily adaptable to hyperbolic systems. The Taylor series of equation (5-3) is written about the point $x_{i-\frac{1}{2}}$, t_n for the variable u, and it is also truncated after the second term of the series. The time derivative is evaluated by equation (6-1). The values of the variables $u_{i-\frac{1}{2},n}$ and $v_{i+\frac{1}{2},n}$ are evaluated as the average of the values at the two grid points on either side. These values are used both for evaluating the nonlinear coefficients and in the term $(u - v)$. The finite difference analog of the space derivative is second-order correct and is

$$\left(\frac{\partial u}{\partial x}\right)_{i-\frac{1}{2},n} = \frac{u_{i,n} - u_{i-1,n}}{\Delta x}$$

The resulting relation for $u_{i-\frac{1}{2},n+\frac{1}{2}}$ is

$$u_{i-\frac{1}{2},n+\frac{1}{2}} = u_{i-\frac{1}{2},n} - \frac{\Delta t}{2}\left\{b_1\left(\frac{u_{i,n} - u_{i-1,n}}{\Delta x}\right) + [c_1(u_{i-\frac{1}{2},n}, v_{i+\frac{1}{2},n})](u_{i-\frac{1}{2},n} - v_{i+\frac{1}{2},n})\right\} \tag{6-3}$$

where

$$u_{i-\frac{1}{2},n} = \tfrac{1}{2}(u_{i,n} + u_{i-1,n})$$

and

$$v_{i+\frac{1}{2},n} = \tfrac{1}{2}(v_{i+1,n} + v_{i,n})$$

The value of $v_{i+\frac{1}{2},n+\frac{1}{2}}$ is obtained in a similar manner, with equation (6-2) being used to evaluate the time derivative. The corresponding relation is

$$v_{i+\frac{1}{2},n+\frac{1}{2}} = v_{i+\frac{1}{2},n} + \frac{\Delta t}{2}\left\{b_2\left(\frac{v_{i+1,n} - v_{i,n}}{\Delta x}\right) + [c_2(u_{i-\frac{1}{2},n}, v_{i+\frac{1}{2},n})](u_{i-\frac{1}{2},n} - v_{i+\frac{1}{2},n})\right\} \tag{6-4}$$

When these values have been computed from (6-3) and (6-4), the values of the nonlinear coefficients, $c_1(u, v)$ and $c_2(u, v)$, can be computed. These can then be used in the finite difference equations, (3-10) and (3-11), and the values at the new time level, $u_{i,n+1}$ and $v_{i,n+1}$, can be computed.

3. BACKWARD PROJECTION

There is also an implicit or backward projection which can be applied to hyperbolic equations. This method, like that described in the previous section, results in the computation of the dependent variable at the half level in space as well as in time. Fortunately, these are the values which are desired for the computation of the nonlinear coefficients. Furthermore, the ratio of space to time increment will be the proper one for minimization of truncation error in the projection if this ratio has been chosen properly for the complete centered difference time step. The backward projection is the best available for hyperbolic equations, as there is no centered projection. The formulation of the equations for the backward projection is somewhat complex, particularly for coupled equations with split boundary conditions. Furthermore, results indicate that iteration is often required even with the forward projection. Consequently, this method has not been tried experimentally, and the equations are not presented herein.

4. COMPARISON OF METHODS

The forward projection and straight iteration have been compared on compressible flow problems (*14*, *15*). It was found that the use of iteration improved the material balance even when the forward projection method was used. In hyperbolic problems the time derivatives remain large throughout the solution. Also, the time increment is related to the space increment to minimize the truncation error, so that very small time increments cannot be used at the beginning of the solution. Consequently, the forward projection method does not give as good an estimate for hyperbolic systems as it does for parabolic systems, and iteration is required. Watts found that, for smaller space increments and, consequently, smaller time increments, fewer iterations were required for convergence of the nonlinear terms. Only three iterations were required for 80 space increments, whereas five were needed for 40 space increments. The forward projection would undoubtedly be satisfactory for 80 increments in the problem studied by Watts. However, only one iteration would be eliminated in each time step. In general, it is not worthwhile to use smaller increments just so the forward projection method can be used for nonlinear coefficients. For problems in which these coefficients do not vary greatly, the forward projection method is satisfactory, but, as a rule, straight iteration is recommended for quasi-linear hyperbolic equations.

5. APPLICATION TO NONLINEAR ORDINARY EQUATIONS

The methods described in the previous sections can be applied to certain nonlinear, first-order ordinary differential equations. The greatest value of these applications is for the solution of coupled equations with two-point boundary conditions. The application is almost identical to that described for second-order equations. As in that case, each stage of the iteration may be regarded as a time step, with the starting values being the initial conditions. Again, an iteration parameter is used in place of the time increment. At this time, no work has been done to find the best values of the iteration parameter. However a small value of this parameter should ensure convergence, and with the high-speed computers, convergence should be fairly rapid. A value which would minimize truncation under the criterion established in Chapter Three might well be a good one.

EXAMPLE 6-1. ISENTROPIC DISCHARGE OF A PERFECT GAS FROM A DUCT

$$\frac{\partial u}{\partial t} = -e^{av}\frac{\partial v}{\partial x} - u\frac{\partial u}{\partial x}$$

$$\frac{\partial v}{\partial t} = -u\frac{\partial v}{\partial x} - \frac{\partial u}{\partial x}$$

$$\begin{aligned} u(0, t) &= 0 \qquad \text{all } t \\ v(1, t) &= 0 \qquad \text{all } t, \text{ if } u(1, t) \gtrless 0 \\ \left.\begin{aligned} u(x, 0) &= 0 \\ v(x, 0) &= v_0 \end{aligned}\right\} &\qquad \text{all } x \end{aligned}$$

Use the centered difference method. Use straight iteration for $u_{i-1/2,n+1/2}$ and $v_{i+1/2,n+1/2}$. Program to perform a fixed number of iterations per time step.

In order that all equations can be computed in the same form, define $x_i = (i - 1)\,\Delta x$ for u and $x_i = (i - 2)\,\Delta x$ for v. Thus, $1 \leq i \leq R + 1$ for u, and $2 \leq i \leq (R + 2)$ for v, with u_1 and v_{R+2} being given by the boundary conditions. Since $u_1 = v_{R+2} = 0$ for all time for these boundary conditions, the same relation can be used for d_2 and d_{R+1} as for d_i.

The equations for values at the half time level are

$$u^{(k)}_{i-1/2,n+1/2} = \tfrac{1}{4}(u_{i-1,n} + u_{i,n} + u^{(k)}_{i-1,n+1} + u^{(k)}_{i,n+1})$$

and

$$v^{(k)}_{i+1/2,n+1/2} = \tfrac{1}{4}(v_{i,n} + v_{i+1,n} + v^{(k)}_{i,n+1} + v^{(k)}_{i+1,n+1})$$

The finite difference equations are

$$\begin{aligned}(1 - u_{i-1/2,n+1/2})u_{i-1,n+1} &+ (1 + u_{i-1/2,n+1/2})u_{i,n+1} - e^{av_{i+1/2,n+1/2}} \cdot v_{i,n+1} \\ &+ e^{av_{i+1/2,n+1/2}} \cdot v_{i+1,n+1} = (1 + u_{i-1/2,n+1/2})u_{i-1,n} + (1 - u_{i-1/2,n+1/2})u_{i,n} \\ &\qquad - e^{av_{i+1/2,n+1/2}} \cdot v_{i+1,n} + e^{av_{i+1/2,n+1/2}} \cdot v_{i,n}\end{aligned}$$

and

$$-u_{i-1,n+1} + u_{i,n+1} + (1 - u_{i-1/2,n+1/2})v_{i,n+1} + (1 + u_{i-1/2,n+1/2})v_{i+1,n+1}$$
$$= u_{i-1,n} - u_{i,n} + (1 + u_{i-1/2,n+1/2})v_{i,n} + (1 - u_{i-1/2,n+1/2})v_{i+1,n}$$

For this program, it will be necessary to store $u_{i,n}$ and $v_{i,n}$ as well as $u_{i,n+1}$ and $v_{i,n+1}$. In this program, call $u_{i,n+1} = U(I)$, $v_{i,n+1} = V(I)$, $u_{i,n} = UN(I)$, and $v_{i,n} = VN(I)$. Also call $u_{i-1/2,n+1/2} = UH$ and $v_{i+1/2,n+1/2} = VH$ with $UM = 1 - UH$, $UP = 1 + UH$, and $VF = \exp(a \cdot VH)$.

For the bi-tridiagonal set, for $2 \leq i \leq (R + 1)$:

$$a_i^{(1)} = UM;\ a_i^{(2)} = 0; \quad a_i^{(3)} = -1;\ a_i^{(4)} = 0$$
$$b_i^{(1)} = UP;\ b_i^{(2)} = -VF;\ b_i^{(3)} = 1; \quad b_i^{(4)} = UM$$
$$c_i^{(1)} = 0; \quad c_i^{(2)} = VF; \quad c_i^{(3)} = 0; \quad c_i^{(4)} = UP$$

The solution algorithm, then, is

$$\beta_i^{(2)} = -VF - UM \cdot \lambda_{i-1}^{(2)};\ \beta_i^{(4)} = UM + \lambda_{i-1}^{(2)}$$
$$\mu_i = UP \cdot \beta_i^{(4)} - \beta_i^{(2)}$$
$$\lambda_i^{(2)} = (VF \cdot \beta_i^{(4)} - UP \cdot \beta_i^{(2)})/\mu_i;\ \lambda_i^{(4)} = (UP \cdot UP - VF)/\mu_i$$
$$\gamma_i^{(1)} = [\beta_i^{(4)}(d_i^{(1)} - UM \cdot \gamma_{i-1}^{(1)}) - \beta_i^{(2)}(d_i^{(2)} + \gamma_{i-1}^{(1)})]/\mu_i$$
$$\gamma_i^{(2)} = [UP(d_i^{(2)} + \gamma_{i-1}^{(1)}) - d_i^{(1)} + UM \cdot \gamma_{i-1}^{(1)}]/\mu_i$$

For these to apply for $i = 2$, $\lambda_1^{(2)}$, $\lambda_1^{(4)}$, $\gamma_1^{(1)}$, and $\gamma_1^{(2)}$ must be defined as 0.

The nomenclature of Example 3-1 will be used: $\beta_i^{(2)} = BT2$, $\beta_i^{(4)} = BT4$, $\lambda_i^{(2)} = EL2(I)$, $\lambda_i^{(4)} = EL4(I)$, $\mu_i = EM$, $\gamma_i^{(1)} = G1(I)$, and $\gamma_i^{(2)} = G2(I)$.

The back solution is $u_R = \gamma_R^{(1)}$, $v_R = \gamma_R^{(2)}$, and $u_i = \gamma_i^{(1)} - \lambda_i^{(2)}v_{i+1}$, and $v_i = \gamma_i^{(2)} - \lambda_i^{(4)}v_{i+1}$.

The number of time steps printed per page is given in Example 3-1. The constant $a = (C_p/C_v) - 1$. The initial density must be low enough that sonic velocity is not reached at the outlet for the boundary conditions given to apply. This problem was run with $a = 0.4$, $v_0 = 0.741$. The boundary condition at $x = 1$ does not apply after inflow begins. Thus, the program called for 100 time steps, with the solution being stopped as soon as $u_{R+1} < 0$. A preliminary run showed that six iterations were required per time step for the nonlinear terms.

The solution as programmed exhibits an oscillation of u_{R+1} at early time steps. A method of suppressing this oscillation is described in reference (*14*).

This program was run for $a = 0.4$, $v_0 = 0.741$, for 20 space increments. It ran 45 time steps before u_{R+1} became less than zero. With WATFOR compiler on an IBM 7044, the running time for 45 time steps was 49 seconds. The running time for Example 3-1 was 60 seconds for 60 time steps. This program requires approximately six times as much computation per time step. Obviously, most of the time required on the computer is for the print-out.

```
      DIMENSION U(102),V(102),UN(102),EL2(102),EL4(102),G1(102),G2(102),
     1VN(102)
      READ (5,30) A,V0,IR,MM
   30 FORMAT (2E10.4,2I3)
      R=IR
      DT=1./R
      T=0.
      N=0
      M=1
      IRP1=IR+1
      IRP2=IR+2
      DO 1 I=1,IRP1
      U(I)=0.
    1 V(I)=V0
      V(IRP2)=0.
      WRITE (6,20)
      WRITE (6,21) N,T
      WRITE (6,22) (U(I),I=1,IRP1)
      WRITE (6,22) (V(I),I=2,IRP2)
      WRITE (6,23)
   20  FORMAT (1H1)
   21 FORMAT (1H0, I5, F11.4)
   22 FORMAT (1H0, (11F11.4))
   23 FORMAT (1H )
      EL2(1)=0.
      EL4(1)=0.
      G1(1)=0.
      G2(1)=0.
      DO 2 N=1,100
      T=T + DT
      M=M +1
      DO 6 I=1,IRP2
      UN(I)=U(I)
    6 VN(I)=V(I)
      DO 7 K=1,6
      DO 3 I=2,IRP1
      UH=0.25* (UN(I-1)+UN(I) + U(I-1) + U(I))
      UP = 1. + UH
      UM = 1. - UH
      VH=0.25 * (VN(I)+VN(I+1)+V(I)+V(I+1))
      VF=EXP(A*VH)
      D1 = UP*UN(I-1) + UM*UN(I) - VF*VN(I+1) + VF*VN(I)
      D2 = UN(I-1) - UN(I) + UP*VN(I) + UM*VN(I+1)
      BT2= -VF -UM * EL2(I-1)
      BT4= UM + EL2(I-1)
      EM= UP*BT4-BT2
      EL2(I)= (VF*BT4-UP*BT2)/EM
      EL4(I)=(UP*UP-VF)/EM
      G1(I)=(BT4*(D1-UM*G1(I-1))-BT2*(D2+G1(I-1)))/EM
    3 G2(I)=(UP*(D2+G1(I-1))- D1 + UM * G1(I-1))/EM
      U(IRP1)=G1(IRP1)
      V(IRP1)=G2(IRP1)
      DO 4 J=2,IR
      I=IRP2-J
      U(I)=G1(I)-EL2(I)*V(I+1)
    4  V(I)=G2(I)-EL4(I)*V(I+1)
    7  CONTINUE
      WRITE (6,21) N,T
      WRITE (6,22) (U(I),I=1,IRP1)
      WRITE (6,22) (V(I),I=2,IRP2)
      WRITE (6,23)
      IF (M-MM) 8,5,5
    5 WRITE (6,20)
      M=0
    8 IF (U(IRP1)) 9,9,2
    2 CONTINUE
    9 CALL EXIT
      END
```

Chapter Seven

NONLINEAR BOUNDARY CONDITIONS

I. SINGLE PARABOLIC EQUATION

All the second-order methods described in the previous chapters have been applied satisfactorily to all types of linear boundary conditions without any complications in the use of existing solution algorithms. These algorithms can be used on problems with nonlinear boundary conditions with little additional complication and with a negligible increase in computer time. The method of handling nonlinear boundary conditions can be described most easily for a single parabolic equation so that the resulting finite difference equations are tridiagonal and can be solved by the Thomas algorithm.

Consider heat conduction in an insulated rod, which is described by

$$\frac{\partial^2 u}{\partial x^2} = \frac{\partial u}{\partial t} \tag{2-1}$$

Let one end be held at a constant temperature and the other end receive heat by radiation from a constant-temperature source. These boundary conditions are

$$u = u_0 \qquad \text{at } x = 0, \text{ all } t \tag{7-1}$$

$$s(1 - u^4) - \frac{\partial u}{\partial x} = 0 \qquad \text{at } x = 1, \text{ all } t \tag{7-2}$$

with the initial condition

$$u = u_0 \qquad \text{at } t = 0, \text{ all } x \tag{7-3}$$

In this case the temperature has been divided by the absolute temperature of the source of radiation in the normalization process. The parameter s in equation (7-2) is dimensionless and contains the cube of the source temperature, the Stephan-Boltzmann constant, and the thermal conductivity and length of the rod.

The Crank-Nicolson equation can be used for equation (2-1) as shown in Chapter Two, and the boundary condition at $x = 0$ can be handled readily as shown in Chapter One. Consider that the grid points are arranged as shown in Figure 1-2, so that the point x_R is on the boundary at $x = 1$, and a fictitious point, x_{R+1}, is placed one increment beyond the boundary. The backward analog to (2-1) should be used at $x = 1$ to ensure against oscillation of the computed values of u_R. This can be obtained from equation (2-10) as

$$u_{R-1,n+1} + \left[-2 - \frac{(\Delta x)^2}{\Delta t}\right]u_{R,n+1} + u_{R+1,n+1} = \left[-\frac{(\Delta x)^2}{\Delta t}\right]u_{R,n} \quad \text{(2-10a)}$$

The discrete analog to the boundary condition of equation (7-2) is

$$s(1 - u_{R,n+1}^4) - \frac{u_{R+1,n+1} - u_{R-1,n+1}}{2(\Delta x)} = 0 \quad \text{(7-4)}$$

or

$$u_{R+1,n+1} = u_{R-1,n+1} + 2s(\Delta x)(1 - u_{R,n+1}^4) \quad \text{(7-4a)}$$

When the value of $u_{R+1,n+1}$ from equation (7-4a) is substituted into equation (2-10a) as was done with the linear boundary conditions, the resulting equation is nonlinear in $u_{R,n+1}$; this is one of the unknowns to be determined. If this equation were linear, it would have been written as the last equation of the tridiagonal set shown as (1-12) and would be

$$a_R u_{R-1} + b_R u_R = d_R \quad \text{(7-5)}$$

where a_R, b_R, and d_R are known constants. Since this equation is nonlinear, let us write it as

$$a_R u_{R-1} + b_R u_R = h + g u_R^4 \quad \text{(7-6)}$$

where a_R, b_R, h, and g are known constants. Let us, then, write (7-6) as the last equation of the tridiagonal set of finite difference equations. All the rest of these equations are linear. Thus, the first half of the Thomas algorithm, as given in the Appendix, can be applied exactly as shown except for the computation of γ_R, which is also $u_{R,n+1}$. In place of this computation, the coefficients in a nonlinear equation containing only $u_{R,n+1}$ are computed. This equation is

$$u_{R,n+1} = \gamma_R = \frac{h - a_R\gamma_{R-1}}{\beta_R} + \frac{g}{\beta_R} u_{R,n+1}^4 \quad \text{(7-7)}$$

or

$$u_{R,n+1} = p + q u_{R,n+1}^4 \quad \text{(7-7a)}$$

where p and q are known constants defined in equation (7-7). Equation (7-7a) can be solved by any of a number of techniques. Once the value of $u_{R,n+1}$ has been determined, the back half of the Thomas algorithm can be used to obtain the rest of the $u_{i,n+1}$. Thus, the nonlinear boundary condition can be handled with a negligible increase in computer time, as the only iteration required is for the solution of equation (7-7a) which contains a single unknown.

While the time for the computation of $u_{R,n+1}$ from equation (7-7a) is very small for any method of solution, the method of false position, or *regula falsi* method, is a good one to use, since covergence can be assured. For use of this method, let us write equation (7-7) as

$$f(u_{R,n+1}) = p + qu_{R,n+1}^4 - u_{R,n+1} = 0 \tag{7-7b}$$

to define the function $f(u_{R,n+1})$. This function is evaluated at two values of the variable, one on each side of the desired root. Two values which not only bound the root but also lie near it can readily be found for (7-7b). One of these is $u_{R,n}$. The dependent variable is usually monotonic, and it varies almost linearly in small increments of t; therefore, a value which should usually lie on the opposite side of the root from $u_{R,n}$ is $u_{R,n} + 2(u_{R,n} - u_{R,n-1})$. After two values of $u_{R,n+1}$ which lie on opposite sides of the root are found, the next estimate of the root is obtained from

$$u^{(3)} = \frac{f^{(1)}u^{(2)} - f^{(2)}u^{(1)}}{f^{(1)} - f^{(2)}} \tag{7-8}$$

In this relation the superscripts denote the iterative values, and the subscripts are omitted from $u_{R,n+1}$. After $u^{(3)}$ is found, $f^{(3)}$ is obtained from (7-7b). If this value is near enough to zero, the iteration is stopped. If not, $u^{(4)}$ is computed from (7-8) with $u^{(3)}$ and $f^{(3)}$ replacing the pair $u^{(1)}$ and $f^{(1)}$ or $u^{(2)}$ and $f^{(2)}$ which lies on the same side of the root. This procedure can be computed very rapidly, and convergence should be obtained in a few iterations.

2. NONLINEAR BOUNDARY CONDITIONS AT BOTH BOUNDARIES

In order that the method described above can be applied, it is necessary that the space indexing begin at the end at which the boundary condition is linear in the dependent variable. For some problems, the boundary conditions at both ends of the system may be nonlinear relations of the dependent variable. Even this situation can be handled without excessive iteration.

Consider the problem described by equation (2-1) with boundary conditions similar to equation (7-2) applying at both ends, with each end radiating to different temperatures. For this condition, $u_{-1,n+1}$ is related to $u_{0,n+1}$ by a nonlinear relation similar to equation (7-4a), and the finite difference equation for $i = 0$ takes the form

$$b_0u_0 + c_0u_1 = r + su_0^4 \tag{7-9}$$

To begin the solution algorithm at the $i = 0$ end, as in the previous section, an estimate or iterate of $u_{0,n+1}$ must be obtained. One value which could be used is $u_{0,n}$. With this first iterate of $u_{0,n+1}$, the value of $d_0 = r + su_{0,n+1}^4$ is computed, and the computations of the forward half of the algorithm are performed. At this point, the iteration described in the previous section is performed for the boundary conditions at $x = 1$. The back solution of the algorithm is then computed; and, as a result, a new iterate of $u_{0,n+1}$ is obtained. A better value of d_0 can be obtained from this value, and the entire procedure is repeated until two successive values of either $u_{0,n+1}$ or $u_{R,n+1}$ agree to within a prescribed tolerance. This procedure requires the use of the complete solution algorithm for each iteration. A procedure requiring less computation per iteration can be used.

An examination of equations (1-12) reveals that they have exactly the same form if both the variables and the equations are reversed in order. In other words, the algorithm can be used to eliminate all the c coefficients, beginning with the equations for $i = R$, in the forward half of the algorithm. In this way, the first variable computed will be $u_{0,n+1}$, and the rest of the variables can be computed by using the coefficients for increasing values of i in the back solution.

The more efficient procedure for handling nonlinear coefficients at each end of the system should thus be apparent. First, the value of $u_{0,n+1}$ is estimated. The forward half of the algorithm is next computed beginning with $i = 0$. The value of $u_{R,n+1}$ is next computed by using the method of false position. However, the back solution half of the algorithm is not performed. Instead, the forward half of the solution algorithm is computed by use of this value of $u_{R,n+1}$ so that this forward half can be started at the $i = R$ end. The value of $u_{0,n+1}$ is computed from an equation similar to equation (7-7b). The value obtained from this computation is used to start the forward half of the algorithm again from $i = 0$, and a new iterate of $u_{R,n+1}$ is obtained by iteration on equation (7-7b). As soon as the iterates of either $u_{0,n+1}$ or $u_{R,n+1}$ have converged, the back, or solution, half of the algorithm is computed, beginning with the converged value. In this method the back half of the algorithm is performed only once, while only the forward half is computed as a part of each iteration. Furthermore, only a few iterations should be necessary.

3. SET OF SIMULTANEOUS HYPERBOLIC EQUATIONS

Actually, the most important use of nonlinear boundary conditions arises in the flow of compressible fluids. This flow must be described by a set of two or three hyperbolic equations. The treatment of the nonlinear boundary conditions is similar to that for parabolic equations. However, since this application is important, it will be described in some detail. For compressible flow, these nonlinear boundary conditions relate several of the dependent variables in a nonlinear manner. To describe the method of handling these relations, let us consider a particular problem.

The problem to be considered is the discharge of a compressible fluid from a pressurized duct. For the isentropic process the governing equations are

$$\frac{\partial v}{\partial t} = -u \frac{\partial v}{\partial x} - \frac{\partial u}{\partial x} \tag{7-10}$$

$$\frac{\partial u}{\partial t} = -u \frac{\partial u}{\partial x} - e^{av} \frac{\partial v}{\partial x} \tag{7-11}$$

The dependent variable u represents the dimensionless velocity, v is the natural logarithm of the dimensionless density, and a is a constant. Equation (7-10) is a material balance, and equation (7-11) is a force balance. At the closed end of the duct the velocity is zero. At the open end the density is taken as equal to the external density unless the resulting velocity is negative, indicating that the fluid is flowing from the reservoir back into the duct. In this event the boundary condition at the open end is obtained from a force balance between the external reservoir and the end of the duct. In terms of the dimensionless variables, these conditions are

$$u = 0 \qquad \text{at } x = 0 \tag{7-12}$$

and

$$v = 0 \qquad \text{at } x = 1 \tag{7-13}$$

unless the resulting $u(1) < 0$, in which case

$$v = \frac{1}{a} \ln \left(1 - \frac{a}{2} u^2\right) \qquad \text{at } x = 1 \tag{7-14}$$

Equations (7-10) and (7-11) are quasi-linear and can be solved by any of the methods described in previous chapter. The forward projection method results in the least complication in handling the boundary conditions, so its use will be assumed for purposes of this example.

For this method, the $2R$ finite difference equations corresponding to (7-10) and (7-11) can be readily obtained, and all these equations are linear. The velocity at the end where $x = 1$ cannot be eliminated from the equations for $i = R$ to yield a linear equation when condition (7-14) is used; so these equations are

$$a_R^{(1)}u_{R-1,n+1} + b_R^{(1)}u_{R,n+1} + b_R^{(2)}v_{R,n+1} = p + qv_{R+1,n+1} \tag{7-15}$$

$$a_R^{(3)}u_{R-1,n+1} + b_R^{(3)}u_{R,n+1} + b_R^{(4)}v_{R,n+1} = r + sv_{R+1,n+1} \tag{7-16}$$

The terms p, q, r, and s are constants determined from the coefficients and old values. All the rest of the finite difference equations take the form of (4-1) and (4-2). In fact, the only difference between the set of finite difference equations arising from (7-10) and (7-11) and the set of (4-1) and (4-2) is that $d_R^{(1)}$ and $d_R^{(2)}$ in (7-15) and (7-16) are not known but are expressed as functions of $v_{R+1,n+1}$. The algorithm of Appendix C can thus be applied to this system of equations up to the computation of $\gamma_R^{(1)}$, which is also $u_{R,n+1}$. However, this value can be computed by the algorithm as a linear function of $v_{R+1,n+1}$. An examination of the equations of the algorithm will reveal the nature of this function. From the last equations of the forward half of the algorithm.

$$u_{R,n+1} = \frac{[\beta_R^{(4)}(d_R^{(1)} - a_R^{(1)}\gamma_{R-1}^{(1)}) - \beta_R^{(2)}(d_R^{(2)} - a_R^{(3)}\gamma_{R-1}^{(1)})]}{\mu_R} \tag{7-17}$$

The terms $d_R^{(1)}$ and $d_R^{(2)}$ are given as linear functions of $v_{R+1,n+1}$ by the right-hand sides of (7-15) and (7-16). The remaining terms of (7-17) will have been determined by use of the algorithm when the computation has proceeded to the last equation, for which $i = R$, in the forward half of the algorithm. This relation can then be expressed as

$$u_{R,n+1} = h + gv_{R+1,n+1} \tag{7-17a}$$

where h and g are constants.

There are thus two equations relating these variables, the other being (7-14), which specifies that

$$v_{R+1,n+1} = \frac{1}{a}\ln\left(1 - \frac{a}{2}u_{R,n+1}^2\right) \tag{7-14a}$$

It must be remembered that the space index on the variable v is one greater than this index on u at the same point in space; hence the inflow boundary condition on the open end appears as shown above when written in finite difference notation. One of these two variables can be readily eliminated to yield a transcendental equation in a single variable. This equation is

$$f(u_{R,n+1}) = u_{R,n+1} - h - \frac{g}{a}\ln\left(1 - \frac{a}{2}u_{R,n+1}^2\right) = 0 \tag{7-18}$$

The problem has thus been reduced to one of finding the root of (7-18). The method of false position described in Section 1 is particularly useful for this equation.

With the value of $u_{R,n+1}$ obtained from this procedure, $v_{R+1,n+1}$ can be obtained from (7-14a). This value can then be used to compute $d_R^{(1)}$ and $d_R^{(2)}$, and hence $\gamma_R^{(2)}$ which is also $v_{R,n+1}$. The back solution of the algorithm can then be completed to obtain all the values of the $u_{i,n+1}$ and $v_{i,n+1}$. This procedure has been found to be a rapid and an accurate one (*14*).

EXAMPLE 7-1. HEAT FLOW IN ROD WITH HEAT RECEIVED BY RADIATION AT ONE END

$$\frac{\partial^2 u}{\partial x^2} = \frac{\partial u}{\partial t}$$

$$u(x, 0) = u_0 \qquad \text{all } x$$

$$u(0, t) = u_0 \qquad \text{all } t$$

$$s(1 - u^4) - \frac{\partial u}{\partial x} = 0 \qquad \text{at } x = 1, \text{ all } t$$

Work this problem with same spacing and increments as in Example 2-1: $x_i = i(\Delta x)$; $\Delta x = 0.05$.

The finite difference equations are:
For $2 \leq i \leq 19$:

$$u_{i-1,n+1} + \left[-2 - 2\frac{(\Delta x)^2}{\Delta t}\right]u_{i,n+1} + u_{i+1,n+1} = -u_{i-1,n} - u_{i+1,n} + \left[2 - 2\frac{(\Delta x)^2}{\Delta t}\right]u_{i,n}$$

For $i = 1$:

$$\left[-2 - 2\frac{(\Delta x)^2}{\Delta t}\right]u_{1,n+1} + u_{2,n+1} = -u_{2,n} - 2u_0 + \left[2 - 2\frac{(\Delta x)^2}{\Delta t}\right]u_{1,n}$$

For $i = 20$, the backward difference equation is used to prevent oscillation in the values of u_{20}. The resulting finite difference equation is

$$2u_{19,n+1} + \left[-2 - \frac{(\Delta x)^2}{\Delta t}\right]u_{20,n+1} = \left[-\frac{(\Delta x)^2}{\Delta t}u_{20,n} - 2s(\Delta x)\right] + 2s(\Delta x)u_{20,n+1}^4$$

The calculation of this problem is almost identical to that for Example 2-1 through the first half of the Thomas algorithm and the computation of β_{19} and γ_{19}. At this point

$$\beta_{20} = -2 - \frac{(\Delta x)^2}{\Delta t} - 2/\beta_{19}$$

Then, from

$$u_{20,n+1} = \gamma_{20} = (d_{20} - a_{20}\gamma_{19})/\beta_{20}$$

$$= \left[-\frac{(\Delta x)^2}{\Delta t} u_{20,n} - 2s(\Delta x) - a_{20}\gamma_{19}\right] \Big/ \beta_{20} + [2s(\Delta x)/\beta_{20}]u_{20,n+1}^4$$

$$p = \left[-\frac{(\Delta x)^2}{\Delta t} u_{20,n} - 2s(\Delta x) - a_{20}\gamma_{19}\right] \Big/ \beta_{20}$$

and

$$q = 2s(\Delta x)/\beta_{20}$$

so

$$f = p + qu_{20,n+1}^4 - u_{20,n+1} = 0$$

This equation is solved by the method of false position. The starting values are u_0 and 1, since $u_{20,n+1}$ will always lie between these two values. The solution is obtained so that $f_1 < 0$ and $f_2 > 0$.

Once the value of $u_{20,n+1}$ has been determined, the back half of the algorithm is used to obtain the rest of the $u_{i,n+1}$.

This program (on next page) was run for $u_0 = 0.5$, $s = 0.1$. With WATFOR compiler on an IBM 7044, the running time was 35 seconds for 40 time steps.

```
      DIMENSION U(20),B(20),G(20)
      READ (5,30) S,U0
   30 FORMAT (2E10.4)
      DO 1 I=1,20
    1 U(I)=U0
      DT=0.0025
      T=0.
      N=0
      M=1
      WRITE (6,20)
      WRITE (6,21) N,T,DT
      WRITE (6,22) U0,(U(I),I=1,20)
      WRITE (6,23)
   20 FORMAT (1H1)
   21 FORMAT (1H0,I5,2F11.4)
   22 FORMAT (1H0,(11F11.4))
   23 FORMAT (1H )
      DO 2 N=1,40
      BE=0.0025/DT
      BF=-2.-BE
      BS=0.1*S
      T=T+DT
      M=M+1
      BP= 0.005/DT
      BB= -2. - BP
      BD= 2.-BP
      B(1)=BB
      G(1)=(BD*U(1)-U(2) - 2.*U0)/BB
      DO 3 I=2,19
      B(I)=BB-1./B(I-1)
      D=-U(I-1) +BD*U(I)-U(I+1)
    3 G(I)=(D-G(I-1))/B(I)
      B(20)=BF-2./B(19)
      P=(-BE*U(20)-BS-2.*G(19))/B(20)
      Q=BS/B(20)
      F=P+Q*(U0**4)-U0
      IF (F) 10,11,12
   10 F1=F
      F2=P+Q-1 .
      U1=U0
      U2=1.
      GO TO 13
   12 F2=F
      F1=P+Q-1.
      U2=U0
      U1=1.
   13 U3=(F1*U2-F2*U1)/(F1-F2)
      F3=P+Q*(U3**4)-U3
      IF (F3-0.00001) 14,15,16
   14 IF (F3+0.00001) 17,15,15
   16 F2=F3
      U2=U3
      GO TO 13
   17 F1=F3
      U1=U3
      GO TO 13
   11 U3=U0
   15 U(20) =U3
      DO 4 J=1,19
      I=20-J
    4 U(I)=G(I)-U(I+1)/B(I)
      WRITE (6,21) N,T,DT
      WRITE (6,22) U0,(U(I),I=1,20)
      WRITE (6,23)
      IF (M-10) 2,5,5
    5 WRITE (6,20)
      M=0
    2 DT=1.1*DT
      CALL EXIT
      END
```

Chapter Eight

ELLIPTIC EQUATIONS AND PARABOLIC EQUATIONS IN TWO AND THREE SPACE DIMENSIONS

I. INTRODUCTION

The equations that describe heat conduction or diffusion in two or three space dimensions are either parabolic or elliptic partial differential equations. The parabolic equations arise from the unsteady-state problems, whereas the elliptic equations describe the steady-state problems. The finite difference analogs to the derivatives can be formulated in much the same manner as for problems in one space dimension. However, new techniques for the solution of the resulting algebraic equations must be used. Consequently, a number of innovations have been developed for the solution of problems in several space dimensions.

These new techniques can be learned most readily by studying their application to the parabolic equations which arise from the unsteady-state problems. These techniques can be applied to the solution of the steady-state case by considering each iteration for the steady-state solution as a time step in an unsteady-state solution. Furthermore, most of the additional complications arise with the addition of one more space dimension, so that it will be sufficient to consider the simplest two-dimensional, parabolic partial differential equation. This equation is

$$\frac{\partial^2 u}{\partial x^2} + \frac{\partial^2 u}{\partial y^2} = \frac{\partial u}{\partial t} \tag{8-1}$$

Equation (8-1) is linear, and techniques for solution of linear equations will be developed first. Some boundary conditions are needed, and some appropriate simple conditions are

$$\left.\begin{aligned} &u(0, y, t) = 1 \\ &u(1, y, t) = 0 \\ &u(x, 0, t) = 1 \\ &u(x, 1, t) = 0 \\ &u(x, y, 0) = 0, \qquad x > 0, y > 0 \end{aligned}\right\} \tag{8-2}$$

From (8-2) it can be deduced that the region over which the differential equation, (8-1), applies is a square. For such regularly shaped regions most of the types of boundary conditions discussed for the one-dimensional case can be handled in a similar manner for multiple space dimensions with no complication of the techniques to be discussed. Methods of handling conditions on irregularly shaped boundaries are discussed in a later section.

2. NOMENCLATURE FOR DISCRETE VARIABLES IN SEVERAL SPACE DIMENSIONS

In the one-dimensional problem described in Chapter Two, the dependent variable was evaluated at a row of equally spaced points at each time level. For the problem described by equation (8-1) the dependent variable is evaluated at a complete grid of points at each time level. Figure 8-1 is an

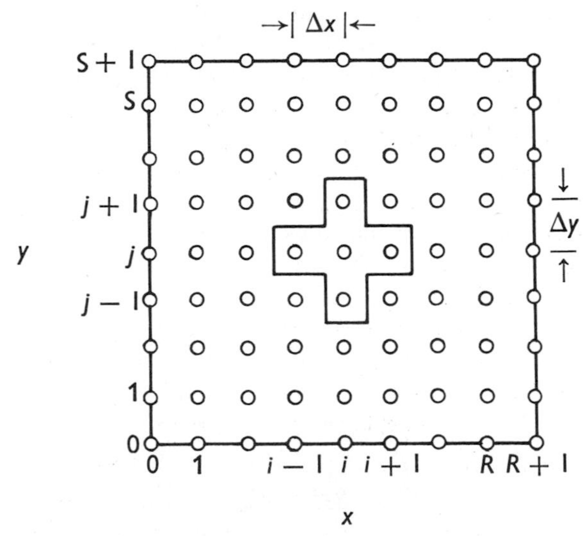

Figure 8-1. Grid points for elliptic equation.

illustration of a grid of points for equation (8-1). Thus, for any two-dimensional problem, the dependent variable is evaluated at a two-dimensional grid at each time level. For a three-dimensional problem the dependent variable is evaluated at a three-dimensional array of points at each time level. The nomenclature for these problems in several space dimensions follows that for problems in a single space dimension. It does get more involved, however.

As an example of the nomenclature for multidimensional problems, consider the problem described by equations (8-1) and (8-2). The location of a discrete point in space and time is specified by three coordinates; thus, a general point is located as x_i, y_j, t_n. The dependent variable is a function of three discrete variables and is denoted as $u(x_i, y_j, t_n)$ or as $u_{i,j,n}$. For problems in three space dimensions and time, an additional subscript is needed. In this case $u(x_i, y_j, z_k, t_n)$ can be denoted as $u_{i,j,k,n}$.

3. CRANK NICOLSON EQUATION

The methods to be used in the solution of multidimensional problems will generally involve values of the dependent variable at two levels in time. The values of the dependent variable at the last known time level, t_n, are used to evaluate the variable at the next time level, t_{n+1}; and the same procedure is used to evaluate the variables at t_{n+2} after all the values at t_{n+1} have been determined. In this respect, the methods for multidimensional problems are similar to the methods developed for problems in one space dimension.

The Crank-Nicolson equation was found to be the most efficient one for the solution of parabolic equations in one space dimension. This equation is thus an obvious one to be used in the solution of parabolic equations in several space dimensions. For the Crank-Nicolson equation, the analogs to the derivatives are centered about the point x_i, y_j, $t_{n+1/2}$. This equation takes the form

$$\frac{1}{(\Delta x)^2}(u_{i+1,j,n+1} + u_{i-1,j,n+1}) + \frac{1}{(\Delta y)^2}(u_{i,j+1,n+1} + u_{i,j-1,n+1}) - 2\left(\frac{1}{(\Delta x)^2} + \frac{1}{(\Delta y)^2} + \frac{1}{(\Delta t)}\right)u_{i,j,n+1} = d_{i,j} \quad (8\text{-}3)$$

In this equation the term $d_{i,j}$ contains the values of the dependent variable at the old time level at the same points in space that appear on the left-hand side. Each finite difference equation contains five unknown values of u. In physical space, these five unknowns are arranged in the pattern of a cross as shown in Figure 8-1.

The finite difference equations (8-3) can be written in matrix form with the components of the unknown vector being written in the order $u_{1,1}$, $u_{2,1}$, $\cdots$, $u_{i,1}$, $\cdots$, $u_{R,1}$, $u_{1,2}$, $\cdots$, $u_{R,2}$, $\cdots$, $u_{1,j}$, $\cdots$, $u_{i,j}$, $\cdots$, $u_{R,j}$, $\cdots$, $u_{1,S}$, $\cdots$, $u_{R,S}$. The five non-zero terms in each row of the coefficient matrix are then arranged as follows: The coefficients of $u_{i-1,j}$, $u_{i,j}$, and $u_{i+1,j}$ are grouped on the main diagonal and the two diagonals just adjacent to it. The coefficients of $u_{i,j-1}$ occupy the diagonal R terms to the left of the main diagonal, and the coefficients of $u_{i,j+1}$ occupy the diagonal R terms to the right of the main diagonal, where $R + 1$ is the number of increments in the x direction.

The set of finite difference equations (8-3) can be considered as S vector equations with S unknown vectors, where $S + 1$ is the number of increments in the y direction. Each of the unknown vectors has R components. The finite difference equations take the form

$$\begin{bmatrix} & \cdots & \\ a_{i,1} & b_{i,1} & c_{i,1} \\ & \cdots & \end{bmatrix}\begin{bmatrix} \cdots \\ u_{i,1} \\ \cdots \end{bmatrix} + \begin{bmatrix} \cdots \\ f_{i,1} \\ \cdots \end{bmatrix}\begin{bmatrix} \cdots \\ u_{i,2} \\ \cdots \end{bmatrix} = \begin{bmatrix} \cdots \\ d_{i,1} \\ \cdots \end{bmatrix}$$

$$\begin{bmatrix} \cdots \\ e_{i,j} \\ \cdots \end{bmatrix}\begin{bmatrix} \cdots \\ u_{i,j-1} \\ \cdots \end{bmatrix} + \begin{bmatrix} & \cdots & \\ a_{i,j} & b_{i,j} & c_{i,j} \\ & \cdots & \end{bmatrix}\begin{bmatrix} \\ u_{i,j} \\ \end{bmatrix} + \begin{bmatrix} \cdots \\ f_{i,j} \\ \cdots \end{bmatrix}\begin{bmatrix} \cdots \\ u_{i,j+1} \\ \cdots \end{bmatrix} = \begin{bmatrix} \cdots \\ d_{i,j} \\ \cdots \end{bmatrix} \quad (8\text{-}3a)$$

$$\begin{bmatrix} \cdots \\ e_{i,S} \\ \cdots \end{bmatrix}\begin{bmatrix} \cdots \\ u_{i,S-1} \\ \cdots \end{bmatrix} + \begin{bmatrix} & \cdots & \\ a_{i,S} & b_{i,S} & c_{i,S} \\ & \cdots & \end{bmatrix}\begin{bmatrix} \cdots \\ u_{i,S} \\ \cdots \end{bmatrix} = \begin{bmatrix} \cdots \\ d_{i,S} \\ \cdots \end{bmatrix}$$

These equations form a tridiagonal set of matrix equations, where the matrices on the main diagonal are themselves tridiagonal and the matrices on each side of the main diagonal are diagonal matrices. This set of equations is, therefore, equivalent to the equations from a one-dimensional problem in which there are S dependent variables.

The solution of (8-3a) by a general matrix inversion routine is impractical. The coefficient matrix contains $(RS)^2$ terms, although most of the elements are zero. A number of efficient iterative methods have been developed for solving (8-3), and these will be discussed in a later section. However, one of the most efficient methods for solving (8-1) and (8-2) numerically is based on a different set of finite difference equations which can be shown to be a factorization of the Crank-Nicolson equations. This is known as the alternating-direction-implicit method.

4. ALTERNATING-DIRECTION-IMPLICIT METHODS

In a number of articles (*10*, *16*, *17*) Douglas *et al.* have described their alternating-direction-implicit methods. For the two-space-dimension problem of (8-1) this method involves the alternate use of two different finite difference analogs to (8-1). For the first finite difference equation the analog to $\partial^2 u/\partial x^2$ is written at the new time level, t_{n+1}, and the analog to the y-derivative,

$\partial^2 u/\partial y^2$, is written at the old time level, t_n. The resulting finite difference equation is

$$\frac{u_{i+1,j,n+1} - 2u_{i,j,n+1} + u_{i-1,j,n+1}}{(\Delta x)^2} + \frac{u_{i,j+1,n} - 2u_{i,j,n} + u_{i,j-1,n}}{(\Delta y)^2} = \frac{u_{i,j,n+1} - u_{i,j,n}}{(\Delta t)} \quad (8\text{-}4)$$

This equation contains three values of the dependent variable at the unknown time level; these are arranged along the horizontal piece of the cross shown in Figure 8-1. It also contains three values of the dependent variable at the known time level, and these are arranged along the vertical piece of the cross. These equations are written with the backward analog in the x-direction and the forward analog in the y-direction. Consequently, they are implicit in the x-direction and explicit in the y-direction.

The finite difference equations for any horizontal row on which the value of j, and consequently y_j, is constant form a tridiagonal coefficient matrix; and their solution can be readily obtained by using the Thomas algorithm described in Appendix A. The values of $u_{i,j,n+1}$ can be obtained by the use of this algorithm S times.

If (8-4) were used continuously, the method would be unstable except for restricted values of the ratio $(\Delta t)/(\Delta y)^2$, since (8-4) is a forward equation in the y-derivative. However, for the next time step, from t_{n+1} to t_{n+2}, (8-4) is not used. Instead, an equation which is explicit in the x-direction and implicit in the y-direction is used with the same size Δt; that is, $t_{n+2} - t_{n+1} = t_{n+1} - t_n$. This equation is

$$\frac{u_{i+1,j,n+1} - 2u_{i,j,n+1} + u_{i-1,j,n+1}}{(\Delta x)^2} + \frac{u_{i,j+1,n+2} - 2u_{i,j,n+2} + u_{i,j-1,n+2}}{(\Delta y)^2} = \frac{u_{i,j,n+2} - u_{i,j,n+1}}{(\Delta t)} \quad (8\text{-}5)$$

Equation (8-5) contains three unknown values of u which are arranged along the vertical piece of the cross in Figure 8-1 and three known values of u which are arranged along the horizontal piece of the cross. The finite difference equations resulting from (8-5) along any vertical column for a constant value of i form a tridiagonal coefficient matrix, so their solution can be obtained from the Thomas algorithm. This algorithm must, therefore, be applied R times to obtain the values of $u_{i,j,n+2}$.

The alternating-direction-implicit method described above has been shown to be stable for any ratio of the time increment to the space increments as long as the same time increment is used for the successive applications of (8-4) and (8-5). The time increment can be increased after any double time step for the solution of problems which approach steady-state conditions. This alternating-direction-implicit method has also been shown to be second-order correct in x, y, and t. See Douglas (*18*) for a complete discussion of the stability and truncation considerations of this method.

The values $u_{i,j,n+1}$ obtained from the application of (8-4) are not representative of the actual values of the dependent variable at time t_{n+1} and should not be used as such. However, if these values are used in (8-5), the values of $u_{i,j,n+2}$ so obtained do approximate very closely the actual values of the dependent variable at time t_{n+2}. For this reason, the values $u_{i,j,n+1}$ should be considered as intermediate values and are sometimes designated as $u^*_{i,j,n+2}$, and the whole double time step is considered as a single time step. This single time step consists of two parts—the computation of the intermediate values from $u_{i,j,n}$ and the computation of the values at the new time level from the intermediate values. In fact, equations (8-4) and (8-5) can be obtained as factors of the Crank-Nicolson equation (8-3). The successive application of (8-4) and (8-5) is, therefore, equivalent to a single application of (8-3). The alternating-direction-implicit methods are preferred over the Crank-Nicolson equation because the resulting finite difference equations can be solved more efficiently.

It may seem implausible that values at one time level, which can be shown to be erroneous, can be used to obtain correct values at the next time level. However, the Crank-Nicolson equation for the one-dimensional, parabolic equation can be shown to be the result of just such a process. This factorization is given in Section 6 of Chapter Two.

5. ALTERNATING-DIRECTION-IMPLICIT METHODS FOR THREE DIMENSIONS

The alternating-direction-implicit methods can also be applied to parabolic equations in three space dimensions. Consider the equation

$$\frac{\partial^2 u}{\partial x^2} + \frac{\partial^2 u}{\partial y^2} + \frac{\partial^2 u}{\partial z^2} = \frac{\partial u}{\partial t} \tag{8-6}$$

For this system there must be three steps of computation for each time step. There are actually two methods which have been proposed for these three-dimensional problems, both of which involve three steps. The one with less truncation error was proposed by Brian (*19*) and is given as the three successive finite difference equations below:

$$\Delta_x^2 u^* + \Delta_y^2 u_n + \Delta_z^2 u_n = \frac{2}{\Delta t}(u^* - u_n) \tag{8-7a}$$

$$\Delta_x^2 u^* + \Delta_y^2 u^{**} + \Delta_z^2 u_n = \frac{2}{\Delta t}(u^{**} - u_n) \tag{8-7b}$$

$$\Delta_x^2 u^* + \Delta_y^2 u^{**} + \Delta_z^2 u_{n+1} = \frac{2}{\Delta t}(u_{n+1} - u^{**}) \tag{8-7c}$$

In these equations the space indices i, j, and k have been omitted from the dependent variables, and

$$\Delta_x^2 u_{i,j,k} = \frac{u_{i+1,j,k} - 2u_{i,j,k} + u_{i-1,j,k}}{(\Delta x)^2}$$

$$\Delta_y^2 u_{i,j,k} = \frac{u_{i,j+1,k} - 2u_{i,j,k} + u_{i,j-1,k}}{(\Delta y)^2}$$

$$\Delta_z^2 u_{i,j,k} = \frac{u_{i,j,k+1} - 2u_{i,j,k} + u_{i,j,k-1}}{(\Delta z)^2}$$

No time subscript is given to the two intermediate values of the dependent variable, as they are computed only as an aid to obtaining the values at the new time level. Each of the finite difference equations is implicit in only one direction, and the resulting equations are a large number of linear, algebraic equations with tridiagonal coefficient matrices.

6. POINT ITERATIVE METHODS OF SOLVING THE CRANK-NICOLSON FINITE DIFFERENCE EQUATIONS

For some types of differential equations the alternating-direction-implicit method has been found to be unstable, so some other technique must be used. The finite difference equations (8-3) which result from applying the Crank-Nicolson method to (8-1) can be solved rather efficiently by some point iterative techniques. Although the matrix is large, $RS \times RS$, it is quite sparse, and iterative techniques are efficient. Young (*20*) has a rather complete discussion of some of these techniques.

The point iterative methods involve computations on one single component equation of (8-3a). Such an equation is

$$e_{i,j}u_{i,j-1,n+1} + a_{i,j}u_{i-1,j,n+1} + b_{i,j}u_{i,j,n+1} + c_{i,j}u_{i+1,j,n+1} + f_{i,j}u_{i,j+1,n+1} = d_{i,j} \tag{8-3b}$$

It is necessary to make a first estimate of $u_{i,j,n+1}$ for all values of i and j, and successively better estimates are obtained. These improved estimates are based on the error which results in the use of the old estimates in (8-3b). A convenient nomenclature makes use of a superscript to designate the number of the estimate; ordinarily $u_{i,j,n+1}^{(m)}$ denotes the last estimate which is known for all values of i and j, and $u_{i,j,n+1}^{(m+1)}$ denotes the new value which is based on the error resulting when $u_{i,j,n+1}^{(m)}$ is substituted into (8-3b).

The simplest point iterative equation is

$$u_{i,j,n+1}^{(m+1)} = \frac{(d_{i,j} - e_{i,j}u_{i,j-1,n+1}^{(m)} - a_{i,j}u_{i-1,j,n+1}^{(m)} - c_{i,j}u_{i+1,j,n+1}^{(m)} - f_{i,j}u_{i,j+1,n+1}^{(m)})}{b_{i,j}} \tag{8-8}$$

This equation is used for all values of i and j in any arbitrary but fixed order. This method is known as the Jacobi method. The convergence of the iteration can be speeded considerably by two changes. First, it can be noticed that, for any ordering of the values of i and j, two improved values will be available, generally, for the computation of $u_{i,j,n+1}^{(m+1)}$. Specifically, consider the order that begins with $i = 1$ and $j = 1$, continues with i incremented by 1 until $i = R$, then continues with $i = 1, j = 2$, continues with i being incremented, etc. In this case both $u_{i,j-1,n+1}^{(m+1)}$ and $u_{i-1,j,n+1}^{(m+1)}$ will have been computed at the time of the computation of $u_{i,j,n+1}^{(m+1)}$. If these two values are used in (8-8) in place of the respective old values, the convergence will require approximately half as many iterations as (8-8). This method is called the Gauss-Seidel method.

A further improvement in the speed of iteration can be obtained by including the old value of $u_{i,j,n+1}$ in obtaining the new value for the same point. The general equation for this type of iteration is

$$\begin{aligned} u_{i,j,n+1}^{(m+1)} = u_{i,j,n+1}^{(m)} - p(e_{i,j}u_{i,j-1,n+1}^{(m+1)} + a_{i,j}u_{i-1,j,n+1}^{(m+1)} \\ + b_{i,j}u_{i,j,n+1}^{(m)} + c_{i,j}u_{i+1,j,n+1}^{(m)} + f_{i,j}u_{i,j+1,n+1}^{(m)} - d_{i,j}) \end{aligned} \tag{8-9}$$

This is the method known as the extrapolated Liebmann method. The term p is an iteration parameter which is adjusted to speed the iteration. If $p = (1/b_{i,j})$, equation (8-9) is exactly the Gauss-Seidel method described above. For the determination of the optimum value of this parameter, the reader is referred to Young (*20*).

Any of the point iterative methods are speeded considerably when the first estimate is near to the solution of (8-3b). For all unsteady-state problems a very good estimate is always available, as the value at the old time step, $u_{i,j,n}$, can be used for the first estimate, $u_{i,j,n+1}^{(0)}$. The point iterative methods are consequently relatively rapid for the solution of parabolic equations in two space dimensions.

7. ITERATIVE METHODS FOR SOLVING ELLIPTIC EQUATIONS

The simplest elliptic partial differential equation is

$$\frac{\partial^2 u}{\partial x^2} + \frac{\partial^2 u}{\partial y^2} = 0 \tag{8-10}$$

This equation arises from steady-state conduction problems in two space dimensions. The numerical solution of (8-10) requires that the value of u be determined at each point in a grid such as that shown in Figure 8-1. There is no time variable, so only one value of u is required at each point. The second-order-correct finite difference analog to (8-10) is

$$u_{i-1,j} + u_{i+1,j} + u_{i,j-1} + u_{i,j+1} - 4u_{i,j} = 0 \tag{8-11}$$

when $\Delta x = \Delta y$. This equation is of the form of (8-3b), and the set of finite difference equations representing (8-10) takes the form of (8-3a).

Since the component equations are of the form of (8-3b), any of the point iterative methods can be applied to (8-11) exactly as they are applied to (8-3). The iteration must be carried out until the values converge to the steady-state solution, and it is not necessary to conduct an iteration for a number of time steps. However, there is no way of obtaining as good an initial estimate for the steady-state problem as was available for each time step of the unsteady-state problem. Consequently, the computing effort for the solution to the steady-state problem may not be significantly less than that for the unsteady-state one.

The alternating-direction-implicit methods can be applied to the solution of the elliptic equations in much the same manner that methods for solving quasi-linear parabolic equations in one space dimension were applied to the solution of nonlinear ordinary differential equations. The first estimate can be considered as the initial conditions, with each complete double-step iteration being considered as a double time step of the unsteady-state problem. The increment ratio, $(\Delta x)^2/(\Delta t)$, for the unsteady-state problem can be considered the iteration parameter, ρ, in the steady-state problem. The pair of successive iteration equations take the forms

$$u_{i,j}^{(m+1)} = u_{i,j}^{(m)} + (1/\rho)(u_{i+1,j}^{(m+1)} - 2u_{i,j}^{(m+1)} + u_{i-1,j}^{(m+1)} + u_{i,j-1}^{(m)} - 2u_{i,j}^{(m)} + u_{i,j+1}^{(m)}) \tag{8-12}$$

and

$$u_{i,j}^{(m+2)} = u_{i,j}^{(m+1)} + (1/\rho)(u_{i+1,j}^{(m+1)} - 2u_{i,j}^{(m+1)} + u_{i-1,j}^{(m+1)} + u_{i,j-1}^{(m+2)} - 2u_{i,j}^{(m+2)} + u_{i,j+1}^{(m+2)}) \quad (8\text{-}13)$$

The same value of the iteration parameter must be used for successive applications of (8-12) and (8-13) in a double-step iteration. The values of this iteration parameter for use in successive iterations can be chosen to minimize computing time. Peaceman and Rachford (*10*) give a relation for determining the optimum set of iteration parameters to be used for a given grid size. The alternating-direction-implicit technique has been found to be an efficient one for the numerical solution of elliptic equations.

There are a number of other numerical techniques available for solving elliptic partial differential equations. Some of these are different iteration schemes for solving (8-11) in which the values of u at a block of points are relaxed simultaneously. Other methods are based on different schemes of setting up the finite difference equations. The reader is referred to the literature for a discussion of these other techniques. The methods presented above should be adequate to solve most elliptic equations which arise, and a thorough understanding of these methods should enable the reader to comprehend many of the other methods which appear in the literature.

8. TREATMENT OF NONLINEAR TERMS

The numerical solution of nonlinear elliptic differential equations and parabolic equations in several space variables is much more complicated than the solution of linear equations of the same type. Any type of solution is necessarily an iterative one, and many different methods can be applied. A method proposed by Douglas *et al.* (*4*) has been found to be satisfactory for a number of rather complex problems. This method will be discussed below.

First, it is necessary to consider a nonlinear differential equation. Although the method can be applied to either elliptic or parabolic equations, it will be applied to a parabolic equation, since the discussion will be somewhat more general. Consider the nonlinear equation

$$\frac{\partial}{\partial x}\left(k\frac{\partial u}{\partial x}\right) + \frac{\partial}{\partial y}\left(k\frac{\partial u}{\partial y}\right) = s\left(\frac{\partial u}{\partial t}\right) \quad (8\text{-}14)$$

where $k = k(u)$ and $s = s(u)$. This equation is a simplification of equations which govern the displacement of oil by water from an underground reservoir. The term denoted by k is a conductivity, and that denoted by s is a saturation. Both of these are functions of the dependent variable.

An alternating-direction-implicit method can be used for the solution of this equation with the nonlinear terms being first projected to the $n + \frac{1}{2}$ time level by one of the projection methods described in the previous chapter. An alternate, somewhat more tedious method has been used so that the saturation term will be evaluated at the $n + 1$ level, thus satisfying the material balance. This latter method will be described in some detail.

It is assumed that the values of the dependent variable at the old time level, $u_{i,j,n}$, are known, and the values at the $n + 1$ level are to be determined from these. An alternating-direction-implicit method is to be used, but more than one iteration per double time step is required. In addition to the finite difference analogs to the three derivatives, an iteration term is introduced into the finite difference equations. The conductivity, k, varies only slightly with the dependent variable, u, and its values do not affect the material balance; therefore, this term is evaluated at the old time level for all iterations. The analogs to the spatial derivatives are written in a form similar to that shown in equation (5-14). The saturation variable is evaluated at the $n + \frac{1}{2}$ time level as the average of $s_{i,j,n}$ and the last known iterative value of $s_{i,j,n+1}$.

There are two iterative equations which must be used alternately. These are

$$\Delta_x(k_n \Delta_x u_{n+1}^{(m+\frac{1}{2})})_{i,j} + \Delta_y(k_n \Delta_y u_{n+1}^{(m)})_{i,j} = s_{i,j,n+\frac{1}{2}}^{(m)}(u_{i,j,n+1}^{(m+\frac{1}{2})} - u_{i,j,n})/\Delta t + H_m(\sum k_n)(u_{i,j,n+1}^{(m+\frac{1}{2})} - u_{i,j,n+1}^{(m)}) \tag{8-15}$$

and

$$\Delta_x(k_n \Delta_x u_{n+1}^{(m+\frac{1}{2})})_{i,j} + \Delta_y(k_n \Delta_y u_{n+1}^{(m+1)})_{i,j} = s_{i,j,n+\frac{1}{2}}^{(m)}(u_{i,j,n+1}^{(m+1)} - u_{i,j,n})/\Delta t + H_m(\sum k_n)(u_{i,j,n+1}^{(m+1)} - u_{i,j,n+1}^{(m+\frac{1}{2})}) \tag{8-16}$$

In these equations the superscript denotes the number of iterations which have been made. The values at the old time level are used as the initial estimates, or $u_{i,j,n+1}^{(0)} = u_{i,j,n}$. The saturation variable is evaluated as

$$s_{i,j,n+\frac{1}{2}}^{(m)} = (s_{i,j,n+1}^{(m)} + s_{i,j,n})/2$$

The analog to the derivative in the x-direction can be expanded to take the form

$$\Delta_x(k_n \Delta_x u_{n+1}^{(m+\frac{1}{2})})_{i,j} = [k_{i+\frac{1}{2},j,n}(u_{i+1,j,n+1}^{(m+\frac{1}{2})} - u_{i,j,n+1}^{(m+\frac{1}{2})}) - k_{i-\frac{1}{2},j,n}(u_{i,j,n+1}^{(m+\frac{1}{2})} - u_{i-1,j,n+1}^{(m+\frac{1}{2})})]/(\Delta x)^2 \tag{8-17}$$

The analog to the derivative in the y-direction can be defined in a similar manner. The values of the conductivity at the half levels in space are evaluated as shown in equations (5-13a,b). The iteration terms include a normalizing factor which is defined as

$$\Sigma k_n = k_{i-\frac{1}{2},j,n} + k_{i+\frac{1}{2},j,n} + k_{i,j-\frac{1}{2},n} + k_{i,j+\frac{1}{2},n} \tag{8-18}$$

The iteration parameter is denoted by H_m. Douglas *et al.* used a cycle of six values of H_m, with $H_1 = 1$, and for each succeeding iteration $H_m = 0.22H_{m-1}$. It was found that four cycles of six iterations each were needed for the first few time steps, but that after that only one cycle of six iterations was needed if the time increment was kept small enough.

This method can be applied to other types of nonlinearities, and certain modifications can be made to fit other equations. It is not presented as a method which will apply to all problems but as an illustration of one of a number of ways of handling nonlinear terms.

9. TREATMENT OF CONDITIONS AT IRREGULAR BOUNDARIES

Complications with boundary conditions arise in multispace dimensional problems which do not occur in one space dimension. In multidimensional problems the boundaries are not necessarily parallel to the space coordinates; thus, in some cases it is not possible to place grid points on the boundary in all rows and columns of points. Consider a region bounded by an ellipse, one-fourth of which is shown in Figure 8-2. For the particular arrangement of points shown there are no points exactly on the boundary, although some points lie quite near to the boundary. The increment sizes could have been chosen so that the grid points along the diameters did lie on the boundary. In the arrangement shown, the boundary along the diameters lies half-way between two grid points.

There are two general types of conditions which often apply at the boundary in elliptic and parabolic problems. For purposes of discussion, consider that the simple elliptic equation (8-10) applies inside the region bounded by the quarter of the ellipse shown in Figure 8-2. At the curved boundary either the value of the dependent variable u or that of its derivative normal to the boundary is generally specified. These two boundary conditions are treated in quite different manners for the numerical solution of (8-10). Consequently, these two types of conditions will be discussed separately, with the specification of the dependent variable on the boundary being treated first.

One of the simplest ways to handle this type of boundary condition is to assign the boundary value of the dependent variable to the grid points which lie nearest to the boundary. In effect the boundary of the region is reshaped so that it conforms to the grid of discrete points. The dotted line in Figure 8-2 indicates how the boundary might be drawn through the points in each row and column which lie nearest to the boundary. The value of the dependent variable at each of these points is set at the value which applies at the nearest point on the curved boundary. The finite difference equation thus does not need to be written about these points joined by the dotted line. It must be written only for the interior points where it takes the form of (8-11). The values of the dependent variable at five points arranged in a cross pattern are

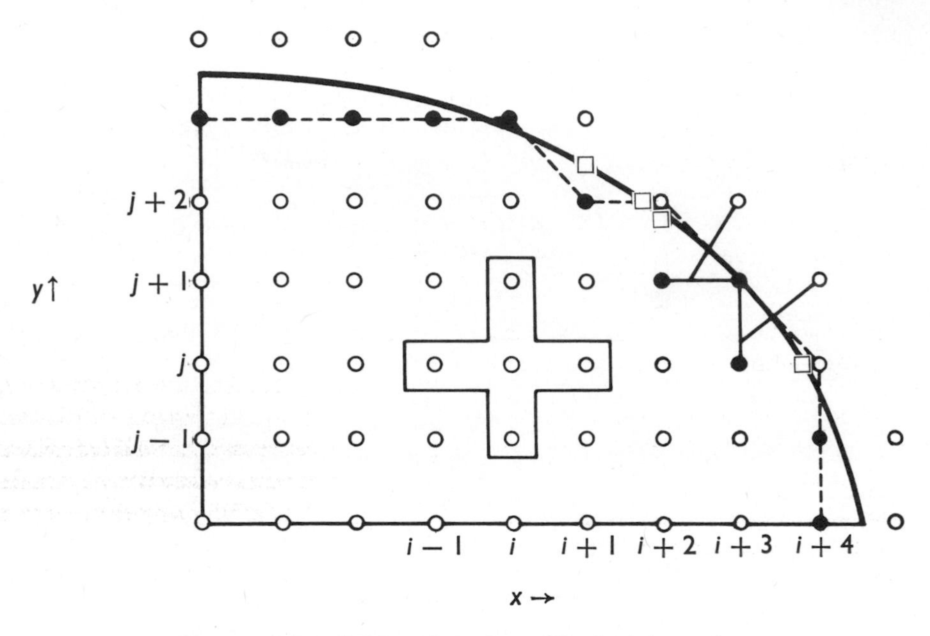

Figure 8-2. Grid points for elliptical boundary.

contained in each equation. This pattern is illustrated in Figure 8-2 for an interior point x_i, y_j, for which all the values of u are unknown. However, for the point x_{i+1}, y_{j+1}, one of the five values, that at x_{i+1}, y_{j+2}, is known by the boundary condition. This method of assigning the boundary conditions thus results in the same number of finite difference equations as there are points at which the value of u is unknown. These equations can be solved by any of the iterative techniques discussed in Section 7 of this chapter. The alternating-direction-implicit method is very satisfactory for this problem. It must be noted, however, that the various rows and columns do not all contain the same number of points, and some complication in the programming of the solution algorithm results.

The procedure just described is a rather simple one for handling the boundary condition in which the value of the dependent variable is specified. However, the finite difference analogs at the points near the boundary are of a rather low-order correctness; as a result, relatively small increments must be used. Another method which results in higher-order-correct analogs can be used. This method is somewhat more trouble to use, but the extra effort is often justified, since larger space increments can be used with the same truncation error. In this method the finite difference equations are written for all interior points, which are divided into two classes. Those interior points which have no neighboring points on the regular grid lying outside the boundary are termed regular points; these are denoted by the open circles inside the boundary in Figure 8-2. Those interior points which have neighboring points on the regular grid lying outside the boundary are termed irregular points; these are denoted by the shaded circles. For the regular interior points, of which x_i, y_j is one, the finite difference equation (8-11) is written.

The finite difference equations for the irregular interior points are altered from this form. Discrete points are placed on the boundary at the intersections with each of the grid lines; a few of these points are shown as squares in Figure 8-2. The values of u at these points are specified by the boundary conditions. For purposes of illustration, consider the finite difference equation for the point (x_{i+2}, y_{j+1}). This point has three interior neighbors (x_{i+3}, y_{j+1}), (x_{i+1}, y_{j+1}), and (x_{i+2}, y_j). Its other neighbor is on the boundary at the intersection with the x_{i+2} grid line. This point is spaced less than a whole increment, Δy, from the point x_{i+2}, y_{j+1}. Let this distance be designated as $s\,\Delta y$, where $0 < s < 1$. The finite difference analog to the $(\partial^2 u/\partial x^2)$ for point x_{i+2}, y_{j+1} is written in the normal manner, since both neighbors along the y_{j+1} grid line are spaced Δx from it. The analog to $(\partial^2 u/\partial y^2)$, however, is written differently, since the two neighbors on the x_{i+2} grid line are unequal distances from it. From the Taylor series the analog to the second derivative for unequally spaced points is

$$\left(\frac{\partial^2 u}{\partial y^2}\right)_{j+1} = \frac{2[u_{j+2} - (s+1)u_{j+1} + su_j]}{s(s+1)(\Delta y)^2} + \frac{(1-s)\,\Delta y}{3}\left(\frac{\partial^3 u}{\partial y^3}\right)_{j+1} - \frac{(s^2 - s + 1)(\Delta y)^2}{12}\left(\frac{\partial^4 u}{\partial y^4}\right)_{j+1} + \cdots + \tag{8-19}$$

This finite difference analog is only first-order correct, although the truncation error becomes second order as the value of s approaches 1. A second-order-correct analog to the first derivative can be written for unequally spaced points; this is

$$\left(\frac{\partial u}{\partial y}\right)_{j+1} = \frac{u_{j+2} - (1 - s^2)u_{j+1} - s^2 u_j}{s(s + 1)\,\Delta y} - \frac{s(\Delta y)^2}{6}\left(\frac{\partial^3 u}{\partial y^3}\right)_{j+1} + \cdots + \tag{8-20}$$

It can be used for elliptic problems which contain first derivatives.

For some of the irregular interior points, such as (x_{i+3}, y_j), only the x-derivative is altered, while for others such as (x_{i+1}, y_{j+2}), both the x-derivative and the y-derivative are altered. With these altered finite difference equations for the irregular interior points, the set of finite difference equations for all interior points is complete. The equations can then be solved by any of the iterative methods described in the previous section.

Two methods have been proposed for handling the boundary condition in which the derivative normal to the boundary is specified. The simpler of these was proposed by Fox (*21*) and is also discussed by Douglas and Peaceman (*22*). In this method a line is drawn from each of the exterior points normal to the boundary. This line is extended until it intersects one of the grid lines joining two of the interior points. In Figure 8-2 this procedure is illustrated for the exterior point (x_{i+3}, y_{j+2}). The line is drawn until it intersects the horizontal grid line for y_{j+1} between points (x_{i+2}, y_{j+1}) and (x_{i+3}, y_{j+1}). The distance between the intersection and point (x_{i+2}, y_{j+1}) is called $r(\Delta x)$, and the distance between the intersection and point (x_{i+3}, y_{j+1}) is called $(1 - r)(\Delta x)$, where $0 < r < 1$. The value of the dependent variable at the intersection is then approximated as the weighted average of the values at the two adjacent grid points. The value of the variable at the exterior point is then related to the values at these two grid points through the boundary condition. As an example, consider the case of the insulated boundary where the derivative normal to the boundary is zero. A first-order-correct finite difference analog to the normal derivative to the boundary results in the value of $u_{i+3,j+2}$ being equal to the value of u at the intersection, or

$$u_{i+3,j+2} = (1 - r)u_{i+2,j+1} + ru_{i+3,j+1} \tag{8-21}$$

In a similar manner, the values of the dependent variable at all exterior points can be expressed as linear functions of values of the dependent variable at interior points. Of course, the normals from some of the exterior grid points intersect a vertical grid line; that from point (x_{i+4}, y_{j+1}) is an example. No complication results in obtaining an expression similar to (8-21) for these points.

For these boundary conditions, the interior points are divided into regular and irregular points just as in the previous illustration. The finite difference equations for the regular interior points will again take the form of (8-11).

For the irregular interior points the finite difference equation can first be written in the same manner including the exterior points, and these points can then be eliminated by relations of the form of (8-21). As an example of this procedure, the finite difference equation for (x_{i+3}, y_{j+1}) becomes

$$(2 - r)u_{i+2,j+1} + (1 + s)u_{i+3,j} - (3 + s - r)u_{i+3,j+1} = 0 \tag{8-22}$$

for the insulated boundary. In this equation $s(\Delta x)$ is the distance between point (x_{i+3}, y_{j+1}) and the intersection of the normal from point (x_{i+4}, y_{j+1}) to the vertical grid line joining the points (x_{i+3}, y_{j+1}) and (x_{i+3}, y_j).

The resulting set of finite difference equations can be solved by any of the iterative methods discussed above. When the alternating-direction-implicit technique is used, it is necessary that the lines always be solved in order of decreasing length. Otherwise, there may be unknown values of u from two lines in a given set of equations.

The other method of handling this type of boundary condition is much more complicated and will not be discussed here. This method was developed by Viswanathan (*23*) and is described also by Fox (*24*). These references can be consulted for a description of this method.

10. CONCLUSION

The methods described in this chapter are by no means all the methods for solving elliptic equations or even necessarily the best methods. Improved methods are being developed and published continually. However, this chapter should enable the reader to solve many problems of this type and to understand and use the newer methods as they are developed.

EXAMPLE 8-1. TWO-DIMENSIONAL STEADY-STATE CONDUCTION

$$\frac{\partial^2 T}{\partial x^2} + \frac{\partial^2 T}{\partial y^2} = 0$$

$$\frac{\partial T}{\partial x} = 0 \qquad \text{at } x = 0, \text{ all } y$$

$$\frac{\partial T}{\partial y} = 0 \qquad \text{at } y = 0, \text{ all } x$$

$$T(1, y) = p(y)$$

$$T(x, 1) = q(x)$$

Define $x_i = (i - 1)\,\Delta x$ and $y_j = (j - 1)\,\Delta y$. Set $\Delta x = \Delta y$.

Use an alternating direction iteration procedure with an iteration parameter, r. Program for R increments in the x-direction and for S increments in the y-direction. A rectangular region can consequently be handled with this program.

The x-implicit equations will be used for the first half of a double iteration.

The finite difference equations are:

For the x-implicit direction:

$$T_{i-1,j}^{(k+1)} + \left(-2 - \frac{1}{r}\right) T_{i,j}^{(k+1)} + T_{i+1,j}^{(k+1)} = -T_{i,j-1}^{(k)} + \left(2 - \frac{1}{r}\right) T_{i,j}^{(k)} - T_{i,j+1}^{(k)}$$

For the y-implicit direction:

$$T_{i,j-1}^{(k+2)} + \left(-2 - \frac{1}{r}\right) T_{i,j}^{(k+2)} + T_{i,j+1}^{(k+2)} = -T_{i-1,j}^{(k+1)} + \left(2 - \frac{1}{r}\right) T_{i,j}^{(k+1)} - T_{i+1,j}^{(k+1)}$$

The equations at the boundaries are altered somewhat.

At $i = 1$, $T_{0,j} = T_{2,j}$ for all j.

At $j = 1$, $T_{i,0} = T_{i,2}$ for all i.

As a result, the coefficient of $T_{2,j}$ and $T_{i,2}$ in these equations is 2. Also, the right side of the equation is modified.

At $i = R$, $T_{R+1,j} = p_j$

At $j = S$, $T_{i,S+1} = q_i$

These terms are transferred to the right sides of the appropriate equations.

The resulting tridiagonal sets are the same for both directions. They are

$$a_i = 1 \quad \text{and} \quad b_i = -2 - \frac{1}{r} \quad \text{all } i$$

$$c_1 = 2 \quad \text{and} \quad c_i = 1 \quad \text{for } 2 \leq i \leq (R - 1)$$

The simplified algorithm is:

$$\beta_1 = b_1,\ \beta_2 = b_2 - 2/\beta_1,\ \beta_i = b_1 - 1/\beta_{i-1} \qquad \text{for } 3 \leq i \leq R$$

$$\gamma_1 = d_1/b_1,\ \gamma_i = (d_i - \gamma_{i-1})/\beta_i \qquad \text{for } 2 \leq i \leq R$$

Then

$$T_R = \gamma_R,\ T_i = \gamma_i - T_{i+1}/\beta_i \qquad \text{for } (R - 1) \geq i \geq 2$$

and

$$T_1 = \gamma_1 - 2T_2/\beta_1$$

In the program, call $E1 = -2 - 1/r$ and $E2 = 2 - 1/r$. Compute all the $d_i = D(I)$ first for each direction. Thus, only one array of $T(I)$ need be stored. Call $\beta_i = B(I)$, and $\gamma_i = G(I)$.

The total number of double iterations is read in as MT, the iterations per print-out as NT.

For the program given (on next two pages), the double iteration is the DO 4 Loop on N. The x-implicit part runs from the DO 5 statement through statement 21. The y-implicit part runs from the DO 11 statement through statement 15.

For the program given, a cycle of nine values of r is given. These are the most efficient values for a 20×20 grid given in reference (*20*).

The boundary conditions in the program are $T(x, 1) = q(x) = 1$ and $T(1, y) = p(y) = 0$. The initial guess is $T_{i,j} = \frac{1}{2}$ everywhere. Convergence to six decimal places was obtained after two cycles of nine double iterations per cycle.

Using WATFOR compiler, running time on an IBM 7044 was 34 seconds for 3 cycles of nine double iterations.

```
      DIMENSION  T(102,102),  R(20),G(102), B(102), D(102,102)
   51 READ (5,2) IR, JS, E, NT, MT
    2 FORMAT (I3, I3, E7.1, I2,I3)
      JP = JS + 1
      IP = IR + 1
      JO = 1
      IO = 1
   19 FORMAT(6H  R = ,I3,6H  S = ,I3,6H  E = ,E9.1)
    7 FORMAT (1H )
   20 FORMAT (6HK N = ,I2)
      IM = IR-1
      JM = JS-1
      R(1) = 0.25155
      R(2) = 0.47492
      R(3) = 0.89666
      R(4) = 1.69291
      R(5) = 3.19623
      R(6) = 6.03450
      R(7) = 11.39320
      R(8) = 21.51045
      R(9) = 40.61191
      DO 1 I = 1,IR
      DO 3 J = 1,JS
    3 T(I,J) = .5
    1 T(I,JP) = 1.
      DO 50 J = 1,JS
   50 T(IP,J) = .0
      T(IP,JP) = .5
    8 FORMAT (1H 11F11.8)
   18 FORMAT (1H )
      NN = 0
      DO 4 N=1,MT
      NN = NN + 1
      E = 1./R(NN)
      E1 = -2.-E
      E2 =  2.-E
      DO 5 J = 2,JS
      DO 6  I = 1,IR
    6 D(I,J) = -T(I,J-1) +E2*T(I,J) -T(I,J+1)
    5 CONTINUE
      DO 12  I = 1,IR
   12 D(I,1)=E2*T(I,1) - 2.*T(I,2)
      DO 30 J = 1,JS
   30 D(IR,J) = D(IR,J) - T(IP,J)
      DO 21J =  1,JS
      B(1) = E1
      G(1) = D(1,J)/B(1)
      B(2) = E1 - 2./B(1)
      G(2) =(D(2,J) - G(1)   )/B(2)
      DO 9 I = 3,IR
      B(I) = E1 - 1./B(I-1)
    9 G(I) = (D(I,J) - G(I-1))/B(I)
      T(IR,J) = G(IR)
      DO 10 L = 2,IM
      I = IP - L
   10 T(I,J) = G(I) - T(I+1,J)/B(I)
   21 T(1,J) = G(1) - 2.*T(2,J)/B(1)
      DO 11 J = 1,JS
```

```
11 D(1,J) = E2*T(1,J) - 2.*T(2,J)
   DO 13 I = 2,IR
   DO 14 J = 1,JS
14 D(I,J) = -T(I-1,J) + E2*T(I,J) - T(I+1,J)
13 CONTINUE
   DO 31 I = 1,IR
31 D(I,JS) = D(I,JS) - T(I,JP)
   DO 15 I = 1,IR
   B(1) = E1
   G(1) = D(I,1)/B(1)
   B(2) = E1 - 2./B(1)
   G(2) =(D(I,2) - G(1))/B(2)
   DO 16 J = 3,JS
   B(J) = E1 - 1./B(J-1)
16 G(J) =(D(I,J) - G(J-1))/B(J)
   T(I,JS) = G(JS)
   DO 17  L = 2,JM
   J = JP - L
17 T(I,J) = G(J) - T(I,J+1)/B(J)
15 T(I,1) = G(1) - 2.*T(I,2)/B(1)
   IF (NN - NT) 4,22,22
22 NN = 0
99 FORMAT (1HK)
   DO 23 JI = 1,JP,JO
   J = JP + 1 - JI
   WRITE (6,18)
23 WRITE (6,8) (T(I,J), I =1,IP,IO)
   WRITE (6,99)
 4 CONTINUE
   GO TO 51
   END
```

Chapter Nine

OTHER TYPES OF EQUATIONS

1. INTRODUCTION

The methods described in the previous chapters should enable the reader to solve many of the problems he may encounter. However, complications arise in some cases which cannot be covered in detail. Furthermore, some problems are described by sets of equations which are not included in the coverage of the previous chapters. In these cases considerable ingenuity must be used to develop a method of numerical solution which is both stable and efficient. Methods which have been used on three problems of this type are discussed in this chapter. These discussions are not presented as complete treatises on these problems, nor are they intended to encompass all types of problems. However, a careful study of these cases should provide the reader with an idea of the many variations which can be used in solving various problems numerically. Thus, he will be better prepared to handle other types of equations which might arise.

2. SHOCK WAVES IN HYPERBOLIC EQUATIONS

Hyperbolic equations arise from physical problems for which dispersion or diffusion is neglected. These equations contain no second derivative with respect to length. In Chapter Three it was noted that a finite discontinuity in the boundary condition was propagated through the system without dispersing. A physical result of this type of system is that finite discontinuities in various of the dependent variables can develop during the course of the solution even when these were not initially present. The most well known of these discontinuities is in compressible flow, where the discontinuities are called shocks. Equations (7-10) and (7-11) describe the discharge of a compressible fluid from a pressurized duct. Shocks develop during the course of this process, and special provisions must be made in the numerical solution of these equations to follow the shocks. The numerical solution of this problem has been investigated by von Rosenberg, Beauchamp, and Watts (*14*, *15*, *25*, *26*).

The centered difference method is used as a basis for this solution. It can be used with no provision for the shock or discontinuity during the early part of the solution, and the solution will exhibit a sharp change in velocity, indicating the presence of the shock. This description is adequate as long as the strength of the shock, or extent of the discontinuity, is small. For strong shocks, however, the discontinuity must be inserted into the solution. A complete description of the method for handling the shock is given in Watts, dissertation (*15*) and in a paper in preparation (*26*).

In solving a problem of this type, one must have some understanding of the physical processes involved as well as a knowledge of the governing equations. For compressible flow the extent of the discontinuity is given by the Rankine-Hugoniot relations. These equations relate the values of the dependent variables on the two sides of the discontinuity or shock. For the problem described by equations (7-10) and (7-11), the shock forms at the closed end of the duct at a time when the velocity at the exit begins to drop from its initial value. Initially, this shock has zero strength; that is, the dependent variables are continuous across the shock. However, this shock begins to move into the duct, and the strength of the shock increases. At this time a pair of moving points are introduced into the grid. These points move through the fixed grid of points to represent the shock. In effect, these points are at the same location, and the variables at one of the two points represent the values of the dependent variables in front of the shock, while those at the other point represent the values behind the shock. The dependent variables at these two moving points are related by the Rankine-Hugoniot relations. These moving points usually do not coincide with a fixed point, so that they divide one square region into two trapezoidal regions. The finite difference equations for equations (7-10) and (7-11) are written for each of these trapezoidal regions as well as for the remaining square regions. Thus, there are the same number of equations and unknowns, and the system is defined. Some of these equations are nonlinear, and the exact location of the shock at each time level is a function of the velocity at that time. Thus, an iterative technique of solution must be used. This iteration is combined with that for the nonlinear coefficients, and only six iterations are required per time step. Excellent results were obtained for this method for 80 space increments. The complete unsteady-state problem was computed in 12 minutes on an IBM 7044 digital computer.

3. EQUATIONS FOR FLUX COMPONENTS IN POTENTIAL FLOW

An alternate method for the solution of the steady-state potential flow problem, described by equation (8-10), is obtained if the problem is expressed in terms of the flux components. When this problem is defined in terms of the y- and z-directions as

$$\frac{\partial^2 u}{\partial y^2} + \frac{\partial^2 u}{\partial z^2} = 0 \tag{8-10a}$$

the flux components are defined as

$$v = -k\frac{\partial u}{\partial y} \tag{9-1}$$

and

$$w = -k\frac{\partial u}{\partial z} \tag{9-2}$$

where k is the conductivity. Two differential equations are required to define the problem in terms of the flux components. These are the continuity equation, which is equivalent to equation (8-10a), and the condition for irrotational flow, which must hold so that the flux components can be defined by equations (9-1) and (9-2). These equations are

$$\frac{\partial v}{\partial y} + \frac{\partial w}{\partial z} = 0 \tag{9-3}$$

and

$$\frac{\partial w}{\partial y} - \frac{\partial v}{\partial z} = 0 \tag{9-4}$$

Both these equations contain only first derivatives, just as do equations (3-7) and (3-8). Therefore, a centered difference approach is the obvious one. The method of solution must be somewhat different, however, as the potential flow problem has split boundary conditions in both directions. Typical boundary conditions define either v or w at all boundaries.

The region in which these equations apply is divided in both directions by grid lines. The first approach to this problem is to determine both v and w at all intersections in the grid. A typical section of the grid is shown in Figure 9-1. Finite difference analogs for equations (9-3) and (9-4) are centered about the center of the square unit (the point marked by the cross), and each finite difference equation contains four values of v and four of w. The analogs obtained are second-order correct. Several iterative methods have been developed for the solution of the resulting equations, and these are described in a paper by von Rosenberg (*27*). This method provides an alternate numerical solution of potential flow problems in two dimensions. However, it has no particular advantages over the methods for solving in terms of the potential as described in Chapter Eight.

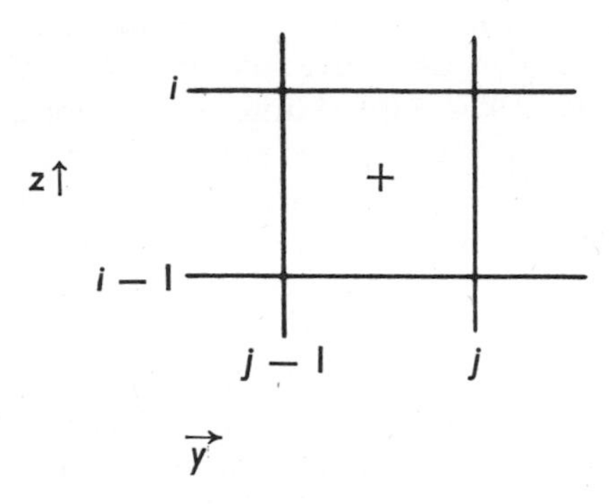

Figure 9-1. Typical unit of grid.

A simple innovation in the method of defining the grid leads to a number of advantages which make the method very promising. This new method is described completely in Gates' dissertation (*28*) and in an article in preparation (*29*). The location for the points of this checkerboard method are shown in Figure 9.2. The dependent variable v is determined at the points designated by the squares (□) in Figure 9-2, and w is determined at those designated by the triangles (△). The analog to equation (9-3), the continuity equation, is centered about the point marked by the open circle (○); and that to equation

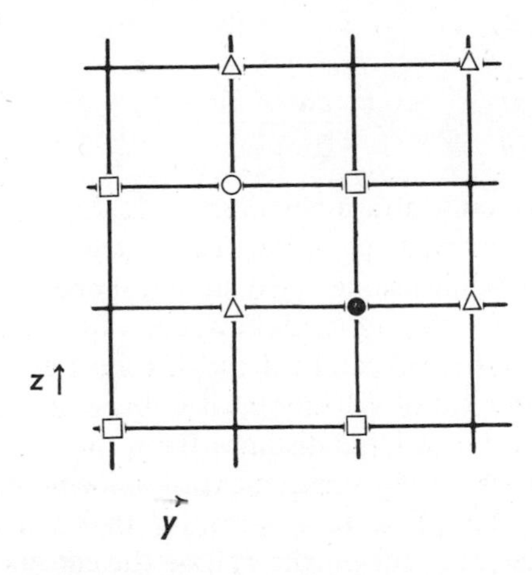

Figure 9-2. Grid for checkerboard method.

(9-4), the irrotationality condition, by the filled circle (●). As a result, each finite difference equation contains only two values of v and two of w. These equations are much simpler, and the equations from adjacent rows can be combined so that all variables can be eliminated except the boundary conditions and the values of v and w from the center rows of points. The coefficient matrix for these equations is of the band form and can be solved by the algorithm of Appendix E. This solution has negligible round-off error for as many as twenty rows of each variable. In this manner, a direct solution is obtained for the potential flow problem with no iteration. Furthermore, even though the checkerboard method is also second-order correct, it has been found to have much less truncation error than the conventional method on several problems for which the exact solution is known. Thus, the innovation of determining variables in a checkerboard pattern has produced several important advantages.

4. BOUNDARY LAYER EQUATIONS

The numerical solution of the unsteady boundary layer equations provides an interesting application of the methods described in the previous chapters. For one dependent variable these equations contain both the first and second derivatives in one space dimension and only first derivatives in the other space dimension and in time. The first derivative with respect to only one of the space dimensions is the only derivative of the other dependent variable. These equations are

$$\frac{\partial u}{\partial t} + u\frac{\partial u}{\partial x} + v\frac{\partial u}{\partial y} = \nu\frac{\partial^2 u}{\partial y^2} \tag{9-5}$$

and

$$\frac{\partial u}{\partial x} + \frac{\partial v}{\partial y} = 0 \tag{9-6}$$

Equation (9-5) is the result of a momentum balance in the x-direction, and equation (9-6) is the continuity equation. The velocity components, u and v, are defined on a two-dimensional grid at a number of time levels. The boundary conditions for this type of flow are often difficult to ascertain. However, a numerical method can be described most easily if a simple set is assumed. The x-component of velocity, u, is defined at two levels of y; these are along the plate and at a large distance from the plate compared to the boundary layer thickness. The y-component of velocity, v, is defined at only one level of y, along the plate. It is necessary that a u velocity profile be specified at some value of x; this might well be the entrance to the boundary layer region. The initial conditions of u must also be specified at all points in the grid.

It is necessary first to define the grid and the locations of the points where u and v are to be determined. A typical grid is shown in Figure 9-3. Values of u are determined at the intersections of the grid lines, and values of v are determined at the points designated by open circles (○). Since the boundary layer thickness is very small, Δy is much smaller than Δx.

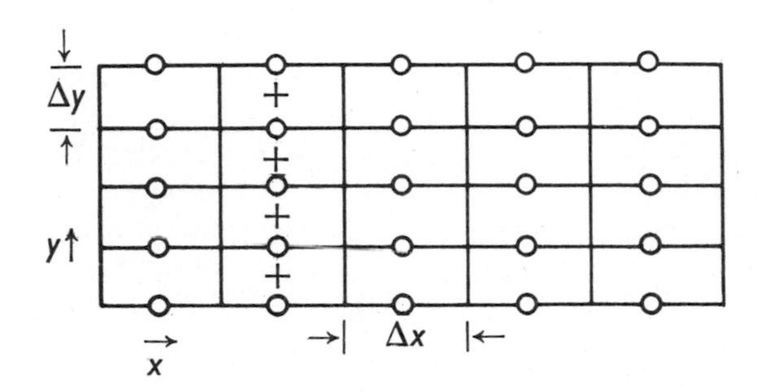

Figure 9-3. Grid for boundary layer problem.

It should be noted that the initial values of v are not required and are not specified by the initial conditions. These can be computed from the initial values of u from equation (9-6), the continuity equation. The analogs to this equation are centered about the center of the rectangle formed by the grid lines. A row of these points is marked by crosses (+). Each finite difference equation contains four values of u and two of v. Of these six values, only the value at the larger y level is unknown, so the computation proceeds explicitly up each vertical column of points, beginning at the plate where the value of v is given by the boundary conditions. Once all the initial values are known, the values of both u and v are computed at the next time level.

The calculation of the values of u at the new time level is made a column at a time from analogs to equation (9-5), with these column-wise calculations being started at the x-boundary where the u velocity profile is known as a function of y from the boundary conditions. The analogs to this equation are centered at the same points in space at which values of v are determined—that is, the points marked with circles (○). However, these analogs are centered in time about the time level half-way between the old and new levels. Thus, these finite difference equations contain values of u at four columns of points; these are columns at two successive values of x at each of the time levels. Only those values at the larger value of x at the new time level are unknown. The resulting equations form a tridiagonal matrix in these values of u, so they can be readily solved by the Thomas algorithm.

Equation (9-5) is quasi-linear, and analogs of u and v must be known at the half time level. The values of u can be projected to the half level by any of the projection methods, and the values of v can be obtained at this level by the analogs to equation (9-6). This method is identical to the method described above for obtaining initial values for v. Once the values of u are obtained at the new time level, values of v at this time level can be obtained, again by the same procedure. This projection method has been used, but a slightly more time-consuming but somewhat simpler iterative method for the quasi-linear coefficients is recommended.

For the iterative method, the values of u at the center points for the finite difference equations are approximated as the average of the four values surrounding it at the same y-level. Likewise, the required value of v is approximated as the average of the values at the two time levels at the same point in space. The first estimate for u and v at the new time level for these nonlinear coefficients is the value at the old time level. With this estimate, values of u at the farther x-level at the new time level are computed. Values of v between the two columns of u at the new time level are next computed from the continuity equation. These values are then used for a new estimate of u and v at the half time level, and the procedure is repeated until the computed values converge. About four iterations seem to be sufficient.

The method described above was used by Killeen (*30*) and subsequently by Stack (*31*). The results obtained were satisfactory, but no experimental data were available to check the numerical solutions. The numerical solution of the steady-state boundary layer equations has since been carried out by a rather obvious simplification of this procedure. The results obtained agree almost exactly with the boundary layer solutions obtained analytically.

REFERENCES

1. Jim Douglas, Jr., "A Survey of Numerical Methods for Parabolic Differential Equations" in *Advances in Computers*, Vol. 2, edited by Franz L. Alt, Academic Press, New York, 1961.

2. H. S. Price, R. S. Varga, and J. E. Warren, *J. Math. Phys.*, *45* (September 1966), 301–311.

3. Burton Wendroff, *J. Ind. Appl. Math.*, *8* (1960), 549.

4. Jim Douglas, Jr., D. W. Peaceman, and H. H. Rachford, Jr., *Petrol. Trans.*, *216* (1959), 297.

5. E. H. Herron, Jr., and D. U. von Rosenberg, *Chem. Eng. Sci.*, *21* (April, 1966), 337–342.

6. P. M. Blair, personal communication.

7. E. H. Herron, Jr., *Distributed Parameter System Dynamics of the Double Pipe Heat Exchanger*, Ph.D. Dissertation in Chemical Engineering, Tulane University, 1964.

8. D. W. Peaceman, personal communication.

9. Jim Douglas, Jr., *Trans. Am. Math. Soc.*, *89* (1958), 484.

10. D. W. Peaceman and H. H. Rachford, Jr., *J. Ind. Appl. Math.*, *3* (1955), 28.

11. D. U. von Rosenberg, P. L. Durrill, and E. H. Spencer, *Brit. Chem. Engr.*, *7* (1962), 186.

12. Thomas R. Harris and Robert E. C. Weaver, *Chem. Eng. Progr. Symp. Ser.*, *62*, No. 66 (1966), 109–122.

13. Jim Douglas, Jr., and B. F. Jones, *J. Ind. Appl. Math.*, *11* (1963), 195.

14. D. U. von Rosenberg, D. L. Beauchamp, and J. W. Watts III, *Chem. Eng. Sci.*, *23* (June, 1968), 345–351.

15. J. W. Watts III, *A Numerical Solution of Transient Compressible Flow from a Duct*, Ph.D. Dissertation in Chemical Engineering, Tulane University, 1967.

16. Jim Douglas, Jr., *Proc. Am. Math. Soc.*, *8* (1957), 409–412.

17. Jim Douglas, Jr., and H. H. Rachford, Jr., *Trans. Am. Math. Soc.*, *82* (1956), 421–439.

18. Jim Douglas, Jr., *J. Soc. Ind. Appl. Math.*, *3* (1955), 42–65.

19. P. L. T. Brian, *Am. Inst. Chem. Eng. J.*, *7* (1961), 367.

20. David Young, Section in *Modern Mathematics for the Engineer*, Second Series, edited by E. F. Beckenbach, McGraw-Hill Book Company, New York, 1961.

21. L. Fox, *Quart. Appl. Math.*, *2* (1944), 251.

22. Jim Douglas, Jr., and D. W. Peaceman, *Am. Inst. Chem. Eng.*, *1* (1955), 505.

23. R. Viswanathan, *Math. Tab.*, *Wash.*, *11* (1957), 67.

24. L. Fox, *Numerical Solution of Ordinary and Partial Differential Equations*, Addison-Wesley Publishing, 1962.

25. D. L. Beauchamp, Jr., *Numerical Solution of Transient Compressible Flow Discharging from a Pipe*, Ph.D. Dissertation in Chemical Engineering, Tulane University, 1966.

26. J. W. Watts III and D. U. von Rosenberg, "A Numerical Solution of a Transient Shock Wave Problem," *Chem. Eng. Sci.*, *24* (January 1969), 49–56.

27. D. U. von Rosenberg, *Math. Comp.*, *21* (1967), 620–28.

28. W. J. Gates, *Numerical Solution for Flux Components in Potential Flow*, Ph.D. Dissertation in Chemical Engineering, Tulane University, 1968.

29. W. J. Gates and D. U. von Rosenberg, "An Improved Numerical Solution for Flux Components in Potential Flow," *Chem. Eng. Sci., 25* (March 1970), 535-547.

30. D. B. Killeen, *The Numerical Solution of Equations Describing an Unsteady Draining Process*, Ph.D. Dissertation in Chemical Engineering, Tulane University, 1966.

31. J. E. Stack III, *The Numerical Solution of a Two-Dimensional Moving Boundary Problem*, Ph.D. Dissertation in Chemical Engineering, Tulane University, 1967.

Appendix A

THOMAS ALGORITHM FOR TRIDIAGONAL MATRIX

The equations are:

$$a_i u_{i-1} + b_i u_i + c_i u_{i+1} = d_i$$
$$\text{for} \quad 1 \leq i \leq R$$
$$\text{with} \quad a_1 = c_R = 0$$

The algorithm is as follows:
First, compute

$$\beta_i = b_i - \frac{a_i c_{i-1}}{\beta_{i-1}} \qquad \text{with } \beta_1 = b_1$$

and

$$\gamma_i = \frac{d_i - a_i \gamma_{i-1}}{\beta_i} \qquad \text{with } \gamma_1 = \frac{d_1}{b_1}$$

The values of the dependent variable are then computed from

$$u_R = \gamma_R \qquad \text{and} \qquad u_i = \gamma_i - \frac{c_i u_{i+1}}{\beta_i}$$

Appendix B

ALGORITHM FOR PENTADIAGONAL MATRIX

The equations are

$$a_i u_{i-2} + b_i u_{i-1} + c_i u_i + d_i u_{i+1} + e_i u_{i+2} = f_i$$

for $1 \leq i \leq R$

with $a_1 = b_1 = a_2 = e_{R-1} = d_R = e_R = 0$

The algorithm is as follows:
First compute

$$\delta_1 = d_1/c_1$$
$$\lambda_1 = e_1/c_1$$
$$\gamma_1 = f_1/c_1$$

and

$$\mu_2 = c_2 - b_2\delta_1$$
$$\delta_2 = (d_2 - b_2\lambda_1)/\mu_2$$
$$\lambda_2 = e_2/\mu_2$$
$$\gamma_2 = (f - b_2\gamma_1)/\mu_2$$

Then for $3 \leq i \leq (R - 2)$, compute

$$\beta_i = b_i - a_i\delta_{i-2}$$
$$\mu_i = c_i - \beta_i\delta_{i-1} - a_i\lambda_{i-2}$$
$$\delta_i = (d_i - \beta_i\lambda_{i-1})/\mu_i$$
$$\lambda_i = e_i/\mu_i$$
$$\gamma_i = (f_i - \beta_i\gamma_{i-1} - a_i\gamma_{i-2})/\mu_i$$

Next compute

$$\beta_{R-1} = b_{R-1} - a_{R-1}\delta_{R-3}$$
$$\mu_{R-1} = c_{R-1} - \beta_{R-1}\delta_{R-2} - a_{R-1}\lambda_{R-3}$$
$$\delta_{R-1} = (d_{R-1} - \beta_{R-1}\lambda_{R-2})/\mu_{R-1}$$
$$\gamma_{R-1} = (f_{R-1} - \beta_{R-1}\gamma_{R-2} - a_{R-1}\gamma_{R-3})/\mu_{R-1}$$

and

$$\begin{aligned}\beta_R &= b_R - a_R\delta_{R-2}\\ \mu_R &= c_R - \beta_R\delta_{R-1} - a_R\lambda_{R-2}\\ \gamma_R &= (f_R - \beta_R\gamma_{R-1} - a_R\gamma_{R-2})/\mu_R\end{aligned}$$

The β_i and μ_i are used only to compute δ_i, λ_i, and γ_i and need not be stored after they are computed. The δ_i, λ_i, and γ_i must be stored, as they are used in the back solution. This is

$$u_R = \gamma_R$$

$$u_{R-1} = \gamma_{R-1} - \delta_{R-1}u_R$$

and

$$u_i = \gamma_i - \delta_i u_{i+1} - \lambda_i u_{i+2}$$

$$\text{for} \quad (R-2) \geq i \geq 1$$

Appendix C

ALGORITHM FOR BI-TRIDIAGONAL MATRIX

The equations are

$$a_i^{(1)}u_{i-1} + a_i^{(2)}v_{i-1} + b_i^{(1)}u_i + b_i^{(2)}v_i + c_i^{(1)}u_{i+1} + c_i^{(2)}v_{i+1} = d_i^{(1)}$$

and

$$a_i^{(3)}u_{i-1} + a_i^{(4)}v_{i-1} + b_i^{(3)}u_i + b_i^{(4)}v_i + c_i^{(3)}u_{i+1} + c_i^{(4)}v_{i+1} = d_i^{(2)}$$

$$\text{for } 1 \le i \le R$$

$$\text{with } a_1^{(m)} = c_R^{(m)} = 0 \qquad \text{for } 1 \le m \le 4$$

The algorithm is as follows:
First compute

$$\beta_i^{(1)} = b_i^{(1)} - a_i^{(1)}\lambda_{i-1}^{(1)} - a_i^{(2)}\lambda_{i-1}^{(3)}$$
$$\beta_i^{(2)} = b_i^{(2)} - a_i^{(1)}\lambda_{i-1}^{(2)} - a_i^{(2)}\lambda_{i-1}^{(4)}$$
$$\beta_i^{(3)} = b_i^{(3)} - a_i^{(3)}\lambda_{i-1}^{(1)} - a_i^{(4)}\lambda_{i-1}^{(3)}$$
$$\beta_i^{(4)} = b_i^{(4)} - a_i^{(3)}\lambda_{i-1}^{(2)} - a_i^{(4)}\lambda_{i-1}^{(4)}$$

$$\text{with } \beta_1^{(m)} = b_1^{(m)} \qquad \text{for } 1 \le m \le 4$$

and

$$\delta_i^{(1)} = d_i^{(1)} - a_i^{(1)}\gamma_{i-1}^{(1)} - a_i^{(2)}\gamma_{i-1}^{(2)}$$
$$\delta_i^{(2)} = d_i^{(2)} - a_i^{(3)}\gamma_{i-1}^{(1)} - a_i^{(4)}\gamma_{i-1}^{(2)}$$

$$\text{with } \delta_1^{(1)} = d_1^{(1)} \qquad \text{and} \qquad \delta_1^{(2)} = d_1^{(2)}$$
$$\text{and} \quad \mu_i = \beta_i^{(1)}\beta_i^{(4)} - \beta_i^{(2)}\beta_i^{(3)}$$

The $\beta_i^{(m)}$, $\delta_i^{(m)}$, and μ_i are computed to aid in the computation of the following functions and need not be stored after the computation of

$$\lambda_i^{(1)} = (\beta_i^{(4)}c_i^{(1)} - \beta_i^{(2)}c_i^{(3)})/\mu_i$$
$$\lambda_i^{(2)} = (\beta_i^{(4)}c_i^{(2)} - \beta_i^{(2)}c_i^{(4)})/\mu_i$$
$$\lambda_i^{(3)} = (\beta_i^{(1)}c_i^{(3)} - \beta_i^{(3)}c_i^{(1)})/\mu_i$$
$$\lambda_i^{(4)} = (\beta_i^{(1)}c_i^{(4)} - \beta_i^{(3)}c_i^{(2)})/\mu_i$$

and

$$\gamma_i^{(1)} = (\beta_i^{(4)}\delta_i^{(1)} - \beta_i^{(2)}\delta_i^{(2)})/\mu_i$$
$$\gamma_i^{(2)} = (\beta_i^{(1)}\delta_i^{(2)} - \beta_i^{(3)}\delta_i^{(1)})/\mu_i$$

The values of $\lambda_i^{(m)}$ and $\gamma_i^{(m)}$ must be stored, as they are used in the back solution. This is

$$u_R = \gamma_R^{(1)}$$
$$v_R = \gamma_R^{(2)}$$

and

$$u_i = \gamma_i^{(1)} - \lambda_i^{(1)}u_{i+1} - \lambda_i^{(2)}v_{i+1}$$
$$v_i = \gamma_i^{(2)} - \lambda_i^{(3)}u_{i+1} - \lambda_i^{(4)}v_{i+1}$$
$$\text{for } (R-1) \geq i \geq 1$$

Appendix D

ALGORITHM FOR TRI-TRIDIAGONAL MATRIX

The equations are

$$a_i^{(1)}u_{i-1} + a_i^{(2)}v_{i-1} + a_i^{(3)}w_{i-1} + b_i^{(1)}u_i + b_i^{(2)}v_i + b_i^{(3)}w_i + c_i^{(1)}u_{i+1} + c_i^{(2)}v_{i+1} + c_i^{(3)}w_{i+1} = d_i^{(1)}$$

$$a_i^{(4)}u_{i-1} + a_i^{(5)}v_{i-1} + a_i^{(6)}w_{i-1} + b_i^{(4)}u_i + b_i^{(5)}v_i + b_i^{(6)}w_i + c_i^{(4)}u_{i+1} + c_i^{(5)}v_{i+1} + c_i^{(6)}w_{i+1} = d_i^{(2)}$$

$$a_i^{(7)}u_{i-1} + a_i^{(8)}v_{i-1} + a_i^{(9)}w_{i-1} + b_i^{(7)}u_i + b_i^{(8)}v_i + b_i^{(9)}w_i + c_i^{(7)}u_{i+1} + c_i^{(8)}v_{i+1} + c_i^{(9)}w_{i+1} = d_i^{(3)}$$

for $1 \leq i \leq R$

with $a_1^{(m)} = c_R^{(m)} = 0 \qquad \text{for } 1 \leq m \leq 9$

The algorithm is as follows:
First compute

$$\beta_i^{(1)} = b_i^{(1)} - a_i^{(1)}\lambda_{i-1}^{(1)} - a_i^{(2)}\lambda_{i-1}^{(4)} - a_i^{(3)}\lambda_{i-1}^{(7)}$$
$$\beta_i^{(2)} = b_i^{(2)} - a_i^{(1)}\lambda_{i-1}^{(2)} - a_i^{(2)}\lambda_{i-1}^{(5)} - a_i^{(3)}\lambda_{i-1}^{(8)}$$
$$\beta_i^{(3)} = b_i^{(3)} - a_i^{(1)}\lambda_{i-1}^{(3)} - a_i^{(2)}\lambda_{i-1}^{(6)} - a_i^{(3)}\lambda_{i-1}^{(9)}$$
$$\beta_i^{(4)} = b_i^{(4)} - a_i^{(4)}\lambda_{i-1}^{(1)} - a_i^{(5)}\lambda_{i-1}^{(4)} - a_i^{(6)}\lambda_{i-1}^{(7)}$$
$$\beta_i^{(5)} = b_i^{(5)} - a_i^{(4)}\lambda_{i-1}^{(2)} - a_i^{(5)}\lambda_{i-1}^{(5)} - a_i^{(6)}\lambda_{i-1}^{(8)}$$
$$\beta_i^{(6)} = b_i^{(6)} - a_i^{(4)}\lambda_{i-1}^{(3)} - a_i^{(5)}\lambda_{i-1}^{(6)} - a_i^{(6)}\lambda_{i-1}^{(9)}$$
$$\beta_i^{(7)} = b_i^{(7)} - a_i^{(7)}\lambda_{i-1}^{(1)} - a_i^{(8)}\lambda_{i-1}^{(4)} - a_i^{(9)}\lambda_{i-1}^{(7)}$$
$$\beta_i^{(8)} = b_i^{(8)} - a_i^{(7)}\lambda_{i-1}^{(2)} - a_i^{(8)}\lambda_{i-1}^{(5)} - a_i^{(9)}\lambda_{i-1}^{(8)}$$
$$\beta_i^{(9)} = b_i^{(9)} - a_i^{(7)}\lambda_{i-1}^{(3)} - a_i^{(8)}\lambda_{i-1}^{(6)} - a_i^{(9)}\lambda_{i-1}^{(9)}$$

with $\beta_1^{(m)} = b_1^{(m)} \qquad \text{for } 1 \leq m \leq 9$

and

$$\delta_i^{(1)} = d_i^{(1)} - a_i^{(1)}\gamma_{i-1}^{(1)} - a_i^{(2)}\gamma_{i-1}^{(2)} - a_i^{(3)}\gamma_{i-1}^{(3)}$$
$$\delta_i^{(2)} = d_i^{(2)} - a_i^{(4)}\gamma_{i-1}^{(1)} - a_i^{(5)}\gamma_{i-1}^{(2)} - a_i^{(6)}\gamma_{i-1}^{(3)}$$
$$\delta_i^{(3)} = d_i^{(3)} - a_i^{(7)}\gamma_{i-1}^{(1)} - a_i^{(8)}\gamma_{i-1}^{(2)} - a_i^{(9)}\gamma_{i-1}^{(3)}$$

with $\delta_1^{(m)} = d_1^{(m)} \qquad \text{for } 1 \leq m \leq 3$

The $\beta_i^{(m)}$ are used to compute several more functions and need not be stored after the computation of

$$\begin{aligned}
\theta_i^{(1)} &= \beta_i^{(5)}\beta_i^{(9)} - \beta_i^{(6)}\beta_i^{(8)} \\
\theta_i^{(2)} &= \beta_i^{(6)}\beta_i^{(7)} - \beta_i^{(4)}\beta_i^{(9)} \\
\theta_i^{(3)} &= \beta_i^{(4)}\beta_i^{(8)} - \beta_i^{(5)}\beta_i^{(7)} \\
\theta_i^{(4)} &= \beta_i^{(3)}\beta_i^{(8)} - \beta_i^{(2)}\beta_i^{(9)} \\
\theta_i^{(5)} &= \beta_i^{(1)}\beta_i^{(9)} - \beta_i^{(3)}\beta_i^{(7)} \\
\theta_i^{(6)} &= \beta_i^{(2)}\beta_i^{(7)} - \beta_i^{(1)}\beta_i^{(8)} \\
\theta_i^{(7)} &= \beta_i^{(2)}\beta_i^{(6)} - \beta_i^{(3)}\beta_i^{(5)} \\
\theta_i^{(8)} &= \beta_i^{(3)}\beta_i^{(4)} - \beta_i^{(1)}\beta_i^{(6)} \\
\theta_i^{(9)} &= \beta_i^{(1)}\beta_i^{(5)} - \beta_i^{(2)}\beta_i^{(4)} \\
\mu_i &= \theta_i^{(1)}\beta_i^{(1)} + \theta_i^{(2)}\beta_i^{(2)} + \theta_i^{(3)}\beta_i^{(3)}
\end{aligned}$$

The $\delta_i^{(m)}$, $\theta_i^{(m)}$, and μ_i are used to compute several more functions and need not be stored after the computation of

$$\begin{aligned}
\lambda_i^{(1)} &= (\theta_i^{(1)}c_i^{(1)} + \theta_i^{(4)}c_i^{(4)} + \theta_i^{(7)}c_i^{(7)})/\mu_i \\
\lambda_i^{(2)} &= (\theta_i^{(1)}c_i^{(2)} + \theta_i^{(4)}c_i^{(5)} + \theta_i^{(7)}c_i^{(8)})/\mu_i \\
\lambda_i^{(3)} &= (\theta_i^{(1)}c_i^{(3)} + \theta_i^{(4)}c_i^{(6)} + \theta_i^{(7)}c_i^{(9)})/\mu_i \\
\lambda_i^{(4)} &= (\theta_i^{(2)}c_i^{(1)} + \theta_i^{(5)}c_i^{(4)} + \theta_i^{(8)}c_i^{(7)})/\mu_i \\
\lambda_i^{(5)} &= (\theta_i^{(2)}c_i^{(2)} + \theta_i^{(5)}c_i^{(5)} + \theta_i^{(8)}c_i^{(8)})/\mu_i \\
\lambda_i^{(6)} &= (\theta_i^{(2)}c_i^{(3)} + \theta_i^{(5)}c_i^{(6)} + \theta_i^{(8)}c_i^{(9)})/\mu_i \\
\lambda_i^{(7)} &= (\theta_i^{(3)}c_i^{(1)} + \theta_i^{(6)}c_i^{(4)} + \theta_i^{(9)}c_i^{(7)})/\mu_i \\
\lambda_i^{(8)} &= (\theta_i^{(3)}c_i^{(2)} + \theta_i^{(6)}c_i^{(5)} + \theta_i^{(9)}c_i^{(8)})/\mu_i \\
\lambda_i^{(9)} &= (\theta_i^{(3)}c_i^{(3)} + \theta_i^{(6)}c_i^{(6)} + \theta_i^{(9)}c_i^{(9)})/\mu_i
\end{aligned}$$

and

$$\begin{aligned}
\gamma_i^{(1)} &= (\theta_i^{(1)}\delta_i^{(1)} + \theta_i^{(4)}\delta_i^{(2)} + \theta_i^{(7)}\delta_i^{(3)})/\mu_i \\
\gamma_i^{(2)} &= (\theta_i^{(2)}\delta_i^{(1)} + \theta_i^{(5)}\delta_i^{(2)} + \theta_i^{(8)}\delta_i^{(3)})/\mu_i \\
\gamma_i^{(3)} &= (\theta_i^{(3)}\delta_i^{(1)} + \theta_i^{(6)}\delta_i^{(2)} + \theta_i^{(9)}\delta_i^{(3)})/\mu_i
\end{aligned}$$

The values of $\lambda_i^{(m)}$ and $\gamma_i^{(m)}$ must be stored, as they are used in the back solution. This is

$$\begin{aligned}
u_R &= \gamma_R^{(1)} \\
v_R &= \gamma_R^{(2)} \\
w_R &= \gamma_R^{(3)}
\end{aligned}$$

and

$$
\begin{aligned}
u_i &= \gamma_i^{(1)} - \lambda_i^{(1)} u_{i+1} - \lambda_i^{(2)} v_{i+1} - \lambda_i^{(3)} w_{i+1} \\
v_i &= \gamma_i^{(2)} - \lambda_i^{(4)} u_{i+1} - \lambda_i^{(5)} v_{i+1} - \lambda^{(6)} w_{i+1} \\
w_i &= \gamma_i^{(3)} - \lambda_i^{(7)} u_{i+1} - \lambda_i^{(8)} v_{i+1} - \lambda_i^{(9)} w_{i+1}
\end{aligned}
$$

$$\text{for } (R-1) \geq i \geq 1$$

Appendix E

GENERAL BAND ALGORITHM

The equations are of the form

$$A_j^{(M)}X_{j-M} + A_j^{(M-1)}X_{j-M+1} + \cdots + A_j^{(2)}X_{j-2} + A_j^{(1)}X_{j-1} + B_jX_j$$
$$C_j^{(1)}X_{j+1} + C_j^{(2)}X_{j+2} + \cdots + C_j^{(M-1)}X_{j+M-1} + C_j^{(M)}X_{j+M} = D_j$$
$$1 \leq j \leq N, \quad N \geq M$$

The algorithm used is as follows:

$$\alpha_j^{(k)} = A_j^{(k)} = 0 \qquad \text{for } k \geq j$$
$$C_j^{(k)} = 0 \qquad \text{for } k \geq N + 1 - j$$

Forward Solution ($j = 1, \cdots, N$):

$$\alpha_j^{(k)} = A_j^{(k)} - \sum_{p=k+1}^{p=M} \alpha_j^{(p)} W_{j-p}^{(p-k)} \qquad k = M, \cdots, 1$$

$$\beta_j = B_j - \sum_{p=1}^{M} \alpha_j^{(p)} W_{j-p}^{(p)}$$

$$W_j^{(k)} = \left(C_j^{(k)} - \sum_{p=k+1}^{p=M} \alpha_j^{(p-k)} W_{j-(p-k)}^{(p)} \right) \Big/ \beta_j \qquad k = 1, \cdots, M$$

$$\gamma_j = \left(D_j - \sum_{p=1}^{M} \alpha_j^{(p)} \gamma_{j-p} \right) \Big/ \beta_j$$

Back Solution ($j = N, \cdots, 1$):

$$X_j = \gamma_j - \sum_{p=1}^{M} W_j^{(p)} X_{j+p}$$

Appendix F

CONVERSION BETWEEN GENERAL TRIDIAGONAL AND BAND FORMS*

I. CONVERSION OF TRIDIAGONAL COEFFICIENTS TO BAND

```
      M = IS + IS - 1
      LS = -IS
      DO 1 L = 1,IS
      LS = IS + LS
      J = -IS + L
      DO 1 I = 1,IR
      J = J + IS
      IF (L - 1) 2,2,3
    3 IF (L - IS) 2,4,4
    2 ISL = IS + L
      DO 5 LL = ISL,M
    5 AA(J,LL) = 0.
      IF (L - 1) 6,6,4
    4 ISL = M - L + 2
      DO 7 LL = ISL,M
    7 CC(J,LL) = 0.
    6 DD(J) = D(I,L)
      DO 1 K = 1,IS
      KS = LS + K
      KL = K - L
      ISP = IS + KL
      ISM = IS - KL
      AA(J,ISM) = A(I,KS)
      IF (KL) 8,9,10
    8 LK = -KL
      AA(J,LK) = B(I,KS)
      GO TO 1
```

* See Section 4-6 for nomenclature

```
  9 BB(J) = B(I,KS)
    GO TO 1
 10 CC(J,KL) = B(I,KS)
  1 CC(J,ISP) = C(I,KS)
```

2. CONVERSION OF BAND SOLUTION VECTOR TO TRIDIAGONAL

```
    JS = -IS
    DO 11 I = 1,IR
    JS = JS + IS
    DO 11 K = 1,IS
    J = JS + K
 11 U(I,K) = X(J)
```

AUTHOR INDEX

Numbers in parentheses indicate the numbers of the references when these are cited in the text without the name of the author.

Numbers set in *italics* designate those page numbers on which the complete literature citations are given.

SUBJECT INDEX